Algebra I Formula Sheet and Key Points

Quick Study Guide and Test Prep Book for Beginners and Advanced Students + Two Algebra I Practice Tests

Dr. Abolfazl Nazari

Copyright © 2024 Dr. Abolfazl Nazari

PUBLISHED BY EFFORTLESS MATH EDUCATION

EFFORTLESSMATH.COM

All rights reserved. No part of this publication may be reproduced, distributed, or transmitted in any form or by any means, including photocopying, recording, or other electronic or mechanical methods, without the prior written permission of the author, except in the case of brief quotations embodied in critical reviews and certain other noncommercial uses permitted by copyright law, including Section 107 or 108 of the 1976 United States Copyright Act.

Copyright ©2024

Algebra I Formula Sheet and Key Points

2024

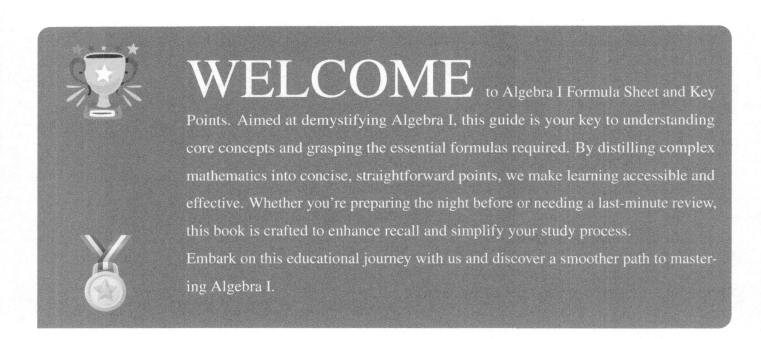

WELCOME to Algebra I Formula Sheet and Key Points. Aimed at demystifying Algebra I, this guide is your key to understanding core concepts and grasping the essential formulas required. By distilling complex mathematics into concise, straightforward points, we make learning accessible and effective. Whether you're preparing the night before or needing a last-minute review, this book is crafted to enhance recall and simplify your study process.
Embark on this educational journey with us and discover a smoother path to mastering Algebra I.

Algebra I Formula Sheet and Key Points is for students eager to learn Algebra I essentials quickly. It distills complex topics down to manageable key points and must-know formulas, facilitating fast learning and retention.

Experience a groundbreaking approach to math learning, designed for quick comprehension and lasting memory of crucial concepts. Understanding the challenge of locating all necessary formulas in one spot, we've dedicated a chapter to compiling all vital formulas needed for the Algebra I examination.

What is included in this book

- ☑ Additional online resources for extended practice and support.
- ☑ A short manual on how to use this book.
- ☑ Coverage of all Algebra I subjects and topics tested.
- ☑ A dedicated chapter listing all the necessary formulas.
- ☑ Two comprehensive, full-length practice tests with thorough answer explanations.

Effortless Math's Algebra I Online Center

Effortless Math Online Algebra I Center offers a complete study program, including the following:

- ☑ *Step-by-step instructions on how to prepare for the Algebra I test*
- ☑ *Numerous Algebra I worksheets to help you measure your math skills*
- ☑ *Complete list of Algebra I formulas*
- ☑ *Video lessons for all Algebra I topics*
- ☑ *Full-length Algebra I practice tests*

Visit EffortlessMath.com/Algebra1 to find your online Algebra I resources.

Scan this QR code

(No Registration Required)

Tips for Making the Most of This Book

This book is your fastest route to mastering Algebra I. It distills the subject down to key points and includes the formulas you need to remember for the Algebra I exam. Here's what makes it special:

- Concise content that focuses on the essentials.
- A dedicated chapter of formulas to remember, covering all topics.
- Two practice tests at the end of the book to assess your knowledge and readiness.

Mathematics can be approachable and easy with the right tools and mindset. Our goal is to simplify math for you, focusing on what's necessary. Using the key points and the formula sheets, you can quickly review and remember the most important information.

Here's how to leverage this book effectively:

Understand and Remember Key Points

Each math topic is built around core ideas or concepts. We've highlighted key points in every topic as mini-summaries of critical information. Don't skip these!

Formulas to Remember

At the end of each topic, we've included a list of formulas to remember. These are the essential formulas you need to know for Algebra I. Make sure to review these regularly.

In summary:

- *Key Points*: Essential summaries of major concepts.
- *Formulas to Remember*: Each topic ends with a list of formulas to remember.

Once you have finished all chapters, review the formula sheet before taking the practice tests. This will help you recall and apply the formulas effectively.

The Practice Tests

The practice tests at the end of the book are a great way to gauge your readiness and identify areas needing more attention. Time yourself and simulate the actual test environment as closely as possible. After completing the practice tests, review your answers to find areas that require additional practice.

Effective Test Preparation

A solid test preparation plan is key. Beyond understanding concepts, strategic study and practice under exam conditions are vital.

- **Begin Early**: Start studying well in advance.
- **Daily Study Sessions**: Regular, short study periods enhance retention.
- **Active Note-Taking**: Helps internalize concepts and improves focus.
- **Review Challenges**: Focus extra time on difficult topics.
- **Practice**: Use end-of-chapter problems and additional resources for extensive practice.

Pick the Right Study Environment and Materials

In addition to this book, consider using other resources for Algebra I. Here are some suggestions: if you need more explanations of the topics, you can use the *Algebra I Made Easy* study guide. For additional practice, try our *Algebra I Workbook*.

Contents

1 Foundations and Basic Principles 1

1.1 Adding and Subtracting Integers 1

1.2 Multiplying and Dividing Integers 1

1.3 Order of Operations 2

1.4 The Distributive Property 2

1.5 Integers and Absolute Value 3

1.6 Proportional Ratios 3

1.7 Similarity and Ratios 3

1.8 Percent Problems 4

1.9 Percent of Increase and Decrease 5

1.10 Discount, Tax, and Tip 5

1.11 Simple Interest 6

1.12 Approximating Irrational Numbers 6

2 Exponents and Variables 7

2.1 Mastering Exponent Multiplication 7

2.2 Exploring Powers of Products 7

2.3	Mastering Exponent Division	8
2.4	Understanding Zero and Negative Exponents	8
2.5	Working with Negative Bases	8
2.6	Introduction to Scientific Notation	9
2.7	Addition and Subtraction in Scientific Notation	9
2.8	Multiplication and Division in Scientific Notation	9
3	**Algebraic Expressions and Equations**	**10**
3.1	Translate a Phrase into an Algebraic Statement	10
3.2	Simplifying Variable Expressions	10
3.3	Evaluating Single Variable Expressions	11
3.4	Evaluating Two Variable Expressions	11
3.5	Solving One-Step Equations	11
3.6	Solving Multi–Step Equations	12
3.7	Rearranging Equations with Multiple Variables	12
3.8	Finding Midpoint	12
3.9	Finding Distance of Two Points	13
4	**Introduction to Linear Functions**	**14**
4.1	Determining Slopes	14
4.2	Formulating Linear Equations	14
4.3	Deriving Equations from Graphs	15
4.4	Understanding Slope-Intercept and Point-Slope Forms	15
4.5	Writing Point-Slope Equations from Graphs	16
4.6	Identifying x- and y-intercepts	16
4.7	Graphing Standard Form Equations	16
4.8	Understanding Horizontal and Vertical Lines	17

4.9	Graphing Horizontal or Vertical Lines	17
4.10	Graphing Point-Slope Form Equations	17
4.11	Understanding Parallel and Perpendicular Lines	17
4.12	Comparing Linear Function Graphs	18
4.13	Graphing Absolute Value Equations	18
4.14	Solving Two-Variable Word Problems	18

5 Inequalities and Systems of Equations 19

5.1	Solving One-Step Inequalities	19
5.2	Solving Multi-Step Inequalities	19
5.3	Working with Compound Inequalities	20
5.4	Graphing Solutions to Linear Inequalities	20
5.5	Writing Linear Inequalities from Graphs	20
5.6	Solving Advanced Linear Inequalities in Two Variables	20
5.7	Graphing Solutions to Advanced Linear Inequalities	21
5.8	Solving Absolute Value Inequalities	21
5.9	Understanding Systems of Equations	21
5.10	Determining the Number of Solutions to Linear Equations	21
5.11	Writing Systems of Equations from Graphs	22
5.12	Solving Systems of Equations Word Problems	22
5.13	Solving Word Problems Involving Linear Equations	22
5.14	Working with Systems of Linear Inequalities	23
5.15	Writing Word Problems for Two-Variable Inequalities	23

6 Polynomial Operations and Functions 24

6.1	Simplifying Polynomial Expressions	24
6.2	Adding and Subtracting Polynomial Expressions	24

6.3	Using Algebra Tiles to Add and Subtract Polynomials	25
6.4	Multiplying Monomial Expressions	25
6.5	Dividing Monomial Expressions	25
6.6	Multiplying a Polynomial by a Monomial	25
6.7	Using Area Models to Multiply Polynomials	26
6.8	Multiplying Binomial Expressions	26
6.9	Using Algebra Tiles to Multiply Binomials	26
6.10	Factoring Trinomial Expressions	26
6.11	Factoring Polynomial Expressions	27
6.12	Using Graphs to Factor Polynomials	27
6.13	Factoring Special Case Polynomial Expressions	28
6.14	Using Polynomials to Find Perimeter	28

7 Quadratic Equations and Functions ... 29

7.1	Solving Quadratic Equations	29
7.2	Graphing Quadratic Functions	30
7.3	Factoring to Solve Quadratic Equations	31
7.4	Understanding Transformations of Quadratic Functions	31
7.5	Completing Function Tables for Quadratic Functions	32
7.6	Determining Domain and Range of Quadratic Functions	32
7.7	Using Algebra Tiles to Factor Quadratics	33
7.8	Writing Quadratic Functions from Vertices and Points	33

8 Relations and Functions ... 34

8.1	Understanding Function Notation and Evaluation	34
8.2	Completing a Function Table from an Equation	34
8.3	Determining Domain and Range of Relations	34

8.4	Performing Addition and Subtraction of Functions	35
8.5	Performing Multiplication and Division of Functions	35
8.6	Composing Functions	35
8.7	Evaluating Exponential Functions	36
8.8	Matching Exponential Functions with Graphs	36
8.9	Writing Exponential Functions from Word Problems	36
8.10	Understanding Function Inverses	37
8.11	Understanding Rate of Change and Slope	37

9	**Radical Expressions and Equations**	**38**
9.1	Simplifying Radical Expressions	38
9.2	Performing Addition and Subtraction of Radical Expressions	38
9.3	Performing Multiplication of Radical Expressions	39
9.4	Rationalizing Radical Expressions	39
9.5	Solving Radical Equations	40
9.6	Determining Domain and Range of Radical Functions	40
9.7	Simplifying Radicals with Fractional Components	41

10	**Rational Expressions**	**42**
10.1	Simplifying Complex Fractions	42
10.2	Graphing Rational Functions	42
10.3	Performing Addition and Subtraction of Rational Expressions	43
10.4	Performing Multiplication of Rational Expressions	43
10.5	Performing Division of Rational Expressions	44
10.6	Evaluating Integers with Rational Exponents	44

11 Number Sequences and Formulas 45

11.1 Evaluating Variable Expressions for Number Sequences 45
11.2 Writing Variable Expressions for Arithmetic Sequences 45
11.3 Writing Variable Expressions for Geometric Sequences 46
11.4 Evaluating Recursive Formulas for Sequences 46
11.5 Formulating a Recursive Sequence 46

12 Direct and Inverse Variations 48

12.1 Determining the Constant of Variation 48
12.2 Formulating and Solving Direct Variation Equations 48
12.3 Modeling Inverse Variation 49

13 Statistical Analysis and Probability 50

13.1 Calculating Mean, Median, Mode, and Range 50
13.2 Creating a Pie Graph 50
13.3 Analyzing Scatter Plots 51
13.4 Calculating and Interpreting Correlation Coefficients 51
13.5 Formulating the Equation of a Regression Line 52
13.6 Understanding Correlation and Causation 52
13.7 Understanding Permutations and Combinations 52
13.8 Solving Probability Problems 53

14 Formula Sheet for All Topics 54

15 Practice Test 1 69

15.1 Practices 69
15.2 Answer Keys 88

15.3	Answers with Explanation	90
16	**Practice Test 2**	**108**
16.1	Practices	108
16.2	Answer Keys	127
16.3	Answers with Explanation	129

1. Foundations and Basic Principles

1.1 Adding and Subtracting Integers

Key Point

In adding or subtracting integers, first observe their signs. If they are alike, add the numbers and maintain the same sign. If different, subtract the smaller from the larger and take the sign of the larger. Also, remember that subtracting a negative is like adding a positive.

Formula To Remember

1. *Adding Two Negative Integers* ☞ $(-a)+(-b) = -(a+b)$ (both a and b are positive).
2. *Adding Positive and Negative Integers* ☞ $a+(-b)$ is equal to $a-b$ if $a > b$, and is equal to $-(b-a)$ if $b > a$, where a and b are positive.
3. *Subtracting Integers Using Addition* ☞ $a-b = a+(-b)$

1.2 Multiplying and Dividing Integers

Key Point

When multiplying or dividing, if both integers are positive or both are negative, the result is positive. But, if one integer is positive and the other negative, the result is negative.

📌 Formula To Remember

1 *Rules for Multiplying Integers* 👉 Let a and b be integers.

$a \times b$ is positive if a and b have the same sign.

$a \times b$ is negative if a and b have opposite signs.

2 *Rules for Dividing Integers* 👉 Let a and b be integers, $b \neq 0$.

$\frac{a}{b}$ is positive if a and b have the same sign.

$\frac{a}{b}$ is negative if a and b have opposite signs.

1.3 Order of Operations

🔔 Key Point

Remember the order of operations with "PEMDAS": First, Parentheses; next, Exponents (like powers and square roots); then, Multiplication and Division (go left to right); and lastly, Addition and Subtraction (also left to right).

📌 Formula To Remember

1 *PEMDAS Rule* 👉 Perform operations in this order: 1. Parentheses, 2. Exponents, 3. Multiplication and Division (from left to right), 4. Addition and Subtraction (from left to right).

1.4 The Distributive Property

🔔 Key Point

The Distributive Property states that multiplying a number by a group of numbers added together is the same as doing each multiplication separately.

📌 Formula To Remember

1 *Distributive Property* 👉 $a \times (b+c) = a \times b + a \times c$

2 *Distributive Property with Subtraction* 👉 $a \times (b-c) = a \times b - a \times c$

1.5 Integers and Absolute Value

🔔 Key Point

The absolute value of an integer is represented as $|a|$, where a is the integer. If a is positive or zero, $|a| = a$, but if a is negative, $|a| = -a$.

📌 Formula To Remember

Definition of Absolute Value 👉 $|a| = \begin{cases} a & \text{if } a \geq 0 \\ -a & \text{if } a < 0. \end{cases}$

1.6 Proportional Ratios

🔔 Key Point

A ratio of 'a' to 'b' implies that for every 'a' units of the first quantity, there are 'b' units of the second quantity.

🔔 Key Point

In a proportion, if $\frac{a}{b} = \frac{c}{d}$, it implies that $ad = bc$ when we cross-multiply. This method, known as cross-multiplication, is a useful technique for solving problems with proportions.

📌 Formula To Remember

Cross-Multiplication 👉 If $\frac{a}{b} = \frac{c}{d}$, then $ad = bc$.

1.7 Similarity and Ratios

🔔 Key Point

Similar figures have proportional corresponding sides and equal corresponding angles. The ratio of corresponding sides is the 'scale factor', which is consistent throughout.

Key Point

In similar figures, the ratio of perimeters is the same as the scale factor. However, the ratio of the areas is the square of the scale factor, while the ratio of volumes is the cube of the scale factor.

Formula To Remember

1. *Scale Factor in Similar Figures* — If two figures are similar with scale factor k, then the ratio of any corresponding linear measurements (sides, perimeters, etc.) is k.
2. *Area Ratio in Similar Figures* — If two figures are similar with scale factor k, the ratio of their areas is k^2.
3. *Volume Ratio in Similar Figures* — If two figures are similar with scale factor k, the ratio of their volumes is k^3.

1.8 Percent Problems

Key Point

A percent represents a fraction with a denominator of 100.

Key Point

The formula to solve percent problems is $b \times \frac{r}{100} = p$. The base ($b$) is what you are taking the percent of. The rate (r) is the percent. The percentage amount (p) is the result of applying the percent to the base.

Formula To Remember

1. *Base Formula for Percent Problems* — $b \times \frac{r}{100} = p$ where b is the base, r is the percent rate, and p is the percentage amount.
2. *Solving for the Base* — $b = \frac{p}{\frac{r}{100}}$ or equivalently $b = \frac{p \times 100}{r}$

1.9 Percent of Increase and Decrease

🔔 Key Point

Percent increase refers to how much the number has gone up as a percentage of the original. Percent decrease, on the contrary, refers to how much the number has reduced as a percentage of the original.

🔔 Key Point

Percent Increase = $\frac{\text{Final Value} - \text{Initial Value}}{\text{Initial Value}} \times 100\%$.

🔔 Key Point

Percent Decrease = $\frac{\text{Initial Value} - \text{Final Value}}{\text{Initial Value}} \times 100\%$.

📌 Formula To Remember

1. *Percent Increase* ☞ Percent Increase = $\frac{\text{Final Value} - \text{Initial Value}}{\text{Initial Value}} \times 100\%$
2. *Percent Decrease* ☞ Percent Decrease = $\frac{\text{Initial Value} - \text{Final Value}}{\text{Initial Value}} \times 100\%$

1.10 Discount, Tax, and Tip

🔔 Key Point

To calculate a discount: Discount = $\frac{\text{Discount Rate}}{100} \times$ Mark Price.

🔔 Key Point

To calculate tax: Tax = $\frac{\text{Tax Rate}}{100} \times$ Price.

🔔 Key Point

To calculate a tip: Tip = $\frac{\text{Tip Rate}}{100} \times$ Total Bill.

Formula To Remember

1. *Discount Calculation* 👉 Discount = $\frac{\text{Discount Rate}}{100} \times$ Mark Price
2. *Tax Calculation* 👉 Tax = $\frac{\text{Tax Rate}}{100} \times$ Price
3. *Tip Calculation* 👉 Tip = $\frac{\text{Tip Rate}}{100} \times$ Total Bill

1.11 Simple Interest

🔔 Key Point

The formula for calculating simple interest is: $I = P \cdot R \cdot T$, where I is the interest, P the principal, R the rate of interest, and T the time in years.

Formula To Remember

1. *Simple Interest Formula* 👉 $I = P \times R \times T$ where I is the interest, P the principal, R the rate of interest, and T the time in years.

1.12 Approximating Irrational Numbers

🔔 Key Point

Numbers that cannot be written as a fraction are called *irrational*. They have non-repeating, non-terminating decimal expansions.

🔔 Key Point

The goal of approximation is to find a value close enough to the true value of an irrational number for practical usage.

2. Exponents and Variables

2.1 Mastering Exponent Multiplication

🔔 Key Point

When multiplying terms with the same base, you simply add the exponents together: $a^n \times a^m = a^{n+m}$. This is known as the *Product of Powers Property*.

📌 Formula To Remember

1. *Product of Powers Property* 👉 $a^n \times a^m = a^{n+m}$

2.2 Exploring Powers of Products

🔔 Key Point

The Power of a Product Property: $(ab)^n = a^n b^n$.

📌 Formula To Remember

1. *Power of a Product Property* 👉 $(ab)^n = a^n b^n$

2.3 Mastering Exponent Division

> **Key Point**
>
> When dividing terms with the same base, subtract the exponents. Therefore, $b^m \div b^n = b^{m-n}$ if $b \neq 0$. This is known as the *Quotient of Powers Property*.

> **Formula To Remember**
>
> 1. *Quotient of Powers Property* 👉 $b^m \div b^n = b^{m-n}$ where $b \neq 0$.

2.4 Understanding Zero and Negative Exponents

> **Key Point**
>
> The Zero Exponent Rule: For any non-zero numeral a, $a^0 = 1$.

> **Key Point**
>
> The Negative Exponent Rule: $a^{-n} = \frac{1}{a^n}$, where a is any non-zero number and n is a positive integer.

> **Formula To Remember**
>
> 1. *Zero Exponent Rule* 👉 $a^0 = 1$ where a is any non-zero number.
> 2. *Negative Exponent Rule* 👉 $a^{-n} = \frac{1}{a^n}$ where a is any non-zero number and n is a positive integer.

2.5 Working with Negative Bases

> **Key Point**
>
> $(-a)^n$ equals a^n if n is an even positive integer, and $-a^n$ if n is an odd positive integer.

> **Formula To Remember**
>
> 1. *Power Rule for Negative Bases* 👉 $(-a)^n = \begin{cases} a^n & \text{if } n \text{ is even} \\ -a^n & \text{if } n \text{ is odd} \end{cases}$

2.6 Introduction to Scientific Notation

🔔 Key Point

Scientific notation involves expressing a number as a product of a number between 1 and 10 and an appropriate power of 10.

📌 Formula To Remember

1. *Scientific Notation Form* $a \times 10^n$ where $1 \leq |a| < 10$ and n is an integer.

2.7 Addition and Subtraction in Scientific Notation

🔔 Key Point

To add or subtract numbers in scientific notation, ensure the exponents for each number are the same.

2.8 Multiplication and Division in Scientific Notation

🔔 Key Point

To multiply numbers in scientific notation, multiply the mantissas and add the exponents. For division, divide the mantissas and subtract the exponents.

📌 Formula To Remember

1. *Multiplying in Scientific Notation* 👉 $(a \times 10^b) \times (c \times 10^d) = (a \times c) \times 10^{b+d}$
2. *Dividing in Scientific Notation* 👉 $\frac{a \times 10^b}{c \times 10^d} = \left(\frac{a}{c}\right) \times 10^{b-d}$

3. Algebraic Expressions and Equations

3.1 Translate a Phrase into an Algebraic Statement

Key Point

The first step in translating a phrase into an algebraic statement is identifying the variables. Next, recognize the operations (addition, subtraction, multiplication, or division) represented by phrases or words.

Formula To Remember

1. *Algebraic Representation of Addition* — Phrase "the sum of a and b" translates to $a + b$.
2. *Algebraic Representation of Multiplication* — Phrase "product of a and b" translates to $a \times b$.
3. *Algebraic Representation of Division* — Phrase "quotient of a and b" translates to $\frac{a}{b}$.
4. *Algebraic Representation of Equality* — Phrase "a is b" translates to $a = b$.

3.2 Simplifying Variable Expressions

Key Point

Always follow the order of operations, commonly expressed as PEMDAS: Parenthesis, Exponents, Multiplication and Division, Addition and Subtraction.

Formula To Remember

1. *Order of Operations - PEMDAS* 👉 Parenthesis, Exponents, Multiplication & Division (left to right), Addition & Subtraction (left to right).

2. *Distributive Property* 👉 $a(b+c) = ab + ac$

3.3 Evaluating Single Variable Expressions

Key Point

To evaluate a single variable expression, substitute the given value of the variable into the expression and simplify.

3.4 Evaluating Two Variable Expressions

Key Point

To evaluate a two-variable expression, replace each variable with its given value and simplify. The order of substitution does not affect the result.

3.5 Solving One-Step Equations

Key Point

For one-step addition or subtraction equations, to get the variable on one side and alone, perform the inverse operation on both sides of the equation.

Key Point

For one-step multiplication or division equations, to get the variable on one side and alone, perform the inverse operation on both sides of the equation.

Formula To Remember

1. *Solving One-Step Addition or Subtraction Equations* — For $x+a=b$, solve by subtracting a from both sides: $x=b-a$.

2. *Solving One-Step Multiplication or Division Equations* — For $ax=b$, solve by dividing both sides by a: $x=\frac{b}{a}$.

3.6 Solving Multi-Step Equations

Key Point

The key to solving multi-step equations is performing one operation at a time, and always keeping the equation balanced by doing the same operation to both sides.

3.7 Rearranging Equations with Multiple Variables

Key Point

The principle of balance is critical when rearranging equations with multiple variables. Any operation performed on one side of the equation must also be performed on the other to maintain equality.

3.8 Finding Midpoint

Key Point

To find the midpoint, M, of a line with endpoints $A(x_1, y_1)$ and $B(x_2, y_2)$, the formula is:

$$M = \left(\frac{x_1 + x_2}{2}, \frac{y_1 + y_2}{2}\right).$$

Formula To Remember

1. *Midpoint Formula* — $M = \left(\frac{x_1+x_2}{2}, \frac{y_1+y_2}{2}\right)$ where (x_1, y_1) and (x_2, y_2) are the coordinates of the endpoints.

3.9 Finding Distance of Two Points

🔔 Key Point

To find the distance between $A(x_1, y_1)$ and $B(x_2, y_2)$, the formula is:

$$d = \sqrt{(x_2 - x_1)^2 + (y_2 - y_1)^2}.$$

🔔 Key Point

The order of points does not change the result. The distance from point A to point B is the same as from point B to point A.

📌 Formula To Remember

1. **Distance Formula** ☞ $d = \sqrt{(x_2 - x_1)^2 + (y_2 - y_1)^2}$

4. Introduction to Linear Functions

4.1 Determining Slopes

Key Point

The slope of a line passing through two points $A(x_1, y_1)$ and $B(x_2, y_2)$ is given by the formula: $m = \frac{y_2 - y_1}{x_2 - x_1}$.

Key Point

Slope characteristics: positive for upward tilt, negative for downward, zero for horizontal, and undefined for vertical lines.

Formula To Remember

1. Slope Formula ☞ $m = \frac{y_2 - y_1}{x_2 - x_1}$ where (x_1, y_1) and (x_2, y_2) are coordinates of two points on the line.

4.2 Formulating Linear Equations

Key Point

A linear equation in slope-intercept form is given by $y = mx + b$ where m is the slope and b is the y-intercept.

Key Point

To find the y-intercept b of a line with a point $P(x_1, y_1)$, use the formula $b = y_1 - mx_1$.

Formula To Remember

1. *Slope-Intercept Form of a Line* 👉 $y = mx + b$ where m is the slope and b is the y-intercept.
2. *Finding the y-Intercept from a Point* 👉 $b = y_1 - mx_1$ where (x_1, y_1) is a point on the line and m is the slope.

4.3 Deriving Equations from Graphs

Key Point

To turn a graph into an equation, we primarily need to identify two aspects – the slope and the y-intercept.

Key Point

The y-intercept is the point where a line crosses the y-axis. In the equation $y = mx + b$, b represents this y-intercept, while m is the slope of the line.

4.4 Understanding Slope-Intercept and Point-Slope Forms

Key Point

The Slope-Intercept Form: Represented as $y = mx + b$, where m is the slope and b is the y-intercept.

Key Point

The Point-Slope Form: Represented as $y - y_1 = m(x - x_1)$, where m is the slope and (x_1, y_1) is a known point on the line.

Formula To Remember

1. *Point-Slope Form of a Line* 👉 $y - y_1 = m(x - x_1)$ where m is the slope and (x_1, y_1) is a known point on the line.

4.5 Writing Point-Slope Equations from Graphs

Key Point

To turn a graph into point-slope equation, identify a specific point and the slope. Plug in these values into the point-slope form $y - y_1 = m(x - x_1)$ to create your equation.

4.6 Identifying x- and y-intercepts

Key Point

The x- and y-intercepts are the points where the line crosses the x-axis and y-axis, respectively. To find the x-intercept, set $y = 0$ in the equation and solve for x. Similarly, to find the y-intercept, set $x = 0$ in the equation and solve for y.

Formula To Remember

1. *Finding the x-intercept* ☞ Set $y = 0$ in the equation of the line and solve for x.
2. *Finding the y-intercept* ☞ Set $x = 0$ in the equation of the line and solve for y.

4.7 Graphing Standard Form Equations

Key Point

The intercept method of graphing involves identifying and marking the x- and y-intercepts on the graph, then joining these points to form the line that represents the equation.

Formula To Remember

1. *Standard Form of a Linear Equation* ☞ $Ax + By = C$ where A, B, and C are constants, with A and B not both zero.

4.8 Understanding Horizontal and Vertical Lines

🔔 Key Point

A key characteristic of horizontal and vertical lines is that their slopes are unique. The slope of a horizontal line is always zero, while a vertical line does not have a finite slope, instead it is considered undefined.

🔔 Key Point

Horizontal lines $y = b$ have no x-intercept, and vertical lines $x = a$ have no y-intercept, except when they pass through the origin.

📌 Formula To Remember

1. Equation of a Horizontal Line 👉 $y = b$ where b is the y-coordinate of all points on the line.
2. Equation of a Vertical Line 👉 $x = a$ where a is the x-coordinate of all points on the line.

4.9 Graphing Horizontal or Vertical Lines

🔔 Key Point

A horizontal line $y = b$, runs parallel to the x-axis, crossing at point b on the y-axis. A vertical line $x = a$, runs parallel to the y-axis, intersecting at point a on the x-axis.

4.10 Graphing Point-Slope Form Equations

🔔 Key Point

For plotting slope-intercept form, we begin by first plotting the known point on the graph. We then use the slope to identify a second point and draw a line through these two points.

4.11 Understanding Parallel and Perpendicular Lines

🔔 Key Point

Two lines are parallel if and only if their slopes are equal.

Chapter 4. Introduction to Linear Functions

Key Point

Two lines are perpendicular if and only if their slopes are negative reciprocals.

Formula To Remember

1 *Condition for Parallel Lines* ☞ Two lines are parallel if their slopes are equal, represented as $m_1 = m_2$, and they do not have the same y-intercept.

2 *Condition for Perpendicular Lines* ☞ Two lines are perpendicular if the slope of one line is the negative reciprocal of the other, expressed as $m_2 = -\frac{1}{m_1}$.

4.12 Comparing Linear Function Graphs

Key Point

A comparison of slopes can show whether lines are parallel, while a comparison of y-intercepts can show where each line crosses the y-axis.

4.13 Graphing Absolute Value Equations

Key Point

The graph of an absolute value equation forms a "V" or inverted "V" shape. The vertex of the graph changes position based on modifications to the equation.

Formula To Remember

1 *Standard Form of Absolute Value Equation* ☞ $y = |x|$ produces a "V" shaped graph with vertex at $(0,0)$.

2 *Effect of Horizontal Shift on Vertex* ☞ $y = |x - h|$ moves the vertex to $(h, 0)$.

4.14 Solving Two-Variable Word Problems

Key Point

The key to solving two-variable word problems lies in identifying the two unknowns, representing them as variables, and setting up the correct equations reflecting the problem's requirements.

5. Inequalities and Systems of Equations

5.1 Solving One-Step Inequalities

Key Point

Keep in mind to change the direction of the inequality sign when you multiply or divide both sides by a negative number. This is a crucial factor while working with inequalities.

Formula To Remember

1. *Solving Addition or Subtraction Inequalities* 👉 $x + c > b$ becomes $x > b - c$; $x - c < b$ becomes $x < b + c$.

2. *Solving Multiplication or Division Inequalities* 👉 $ax < b$ becomes $x < \frac{b}{a}$ and flip the inequality if $a < 0$; $ax > b$ becomes $x > \frac{b}{a}$ and flip the inequality if $a < 0$.

5.2 Solving Multi-Step Inequalities

Key Point

Regardless of the complexity of the inequality, always prioritize isolating the variable. This often simplifies the problem significantly, and you can often break down the problem into several smaller one-step inequalities.

5.3 Working with Compound Inequalities

> **Key Point**
>
> As a rule, *'and'* typically narrows down the solution set (think intersection) while *'or'* tends to expand the solution set (union).

5.4 Graphing Solutions to Linear Inequalities

> **Key Point**
>
> A solid line includes values on the line as solutions, while a dashed line excludes those values.

> **Key Point**
>
> Use $(0,0)$ as the test point (unless it is on the inequality line) and shade the region where this point satisfies the inequality.

5.5 Writing Linear Inequalities from Graphs

> **Key Point**
>
> A dashed line in a graph signifies "$<$" or "$>$", while a solid line signifies "$\leq$" or "$\geq$".

> **Formula To Remember**
>
> 1. *Form of a Linear Inequality* 👉 $y \leq mx + b$ or $y \geq mx + b$ for a solid line, $y < mx + b$ or $y > mx + b$ for a dashed line.

5.6 Solving Advanced Linear Inequalities in Two Variables

> **Key Point**
>
> Advanced linear inequalities in two variables are systems of inequalities where solutions make all inequalities true simultaneously.

5.7 Graphing Solutions to Advanced Linear Inequalities

🔔 Key Point

The solution to an advanced linear inequality in two variables is represented as a region on the coordinate plane, marked by multiple lines of inequality.

5.8 Solving Absolute Value Inequalities

🔔 Key Point

Absolute value inequalities bring in a layer of complexity which we resolve by transforming them into a system of inequalities without any absolute value.

📌 Formula To Remember

1. "Greater than" Absolute Value Inequality 👉 $|A| > B$ is solved by $A > B$ or $A < -B$.
2. "Less than" Absolute Value Inequality 👉 $|A| < B$ is solved by $-B < A < B$.

5.9 Understanding Systems of Equations

🔔 Key Point

A solution to a system of equations is a set of variable values that satisfy all the equations in the system simultaneously.

5.10 Determining the Number of Solutions to Linear Equations

🔔 Key Point

There are three potential outcomes when solving systems of linear equations: one unique solution, infinitely many solutions, and no solution.

Key Point

To determine the number of solutions without solving the system, check the coefficients and constants. If the ratios of the coefficients of x, y, and the constants are equal, the system has infinite solutions. If the coefficients are proportionally equal except for the constants, then there is no solution. Otherwise, the system has only one solution.

Formula To Remember

1. *Criteria for Unique Solution* 👉 A system has exactly one solution if $\frac{a_1}{a_2} \neq \frac{b_1}{b_2}$, considering equations of form $a_1 x + b_1 y = c_1$ and $a_2 x + b_2 y = c_2$.

2. *Criteria for No Solution* 👉 A system has no solution if $\frac{a_1}{a_2} = \frac{b_1}{b_2} \neq \frac{c_1}{c_2}$, where equations are in the form $a_1 x + b_1 y = c_1$ and $a_2 x + b_2 y = c_2$.

3. *Criteria for Infinitely Many Solutions* 👉 A system has infinitely many solutions if $\frac{a_1}{a_2} = \frac{b_1}{b_2} = \frac{c_1}{c_2}$ for equations $a_1 x + b_1 y = c_1$ and $a_2 x + b_2 y = c_2$.

5.11 Writing Systems of Equations from Graphs

Key Point

Looking at a graph, one can write down a system of equations by deciding the equations representing the lines and figuring out the intersection points.

5.12 Solving Systems of Equations Word Problems

Key Point

To tackle system of equations word problems, we translate the problem's text into a system of equations, solve it, and then interpret the solution within the context of the problem.

5.13 Solving Word Problems Involving Linear Equations

Key Point

Solving a word problem involving linear equations entails translating the problem into a linear equation, solving the equation, and interpreting the result in the problem's real-world context.

5.14 Working with Systems of Linear Inequalities

Key Point

All feasible solutions to a system of linear inequalities form a solution region on the graph which satisfies all inequalities in the system.

Key Point

Solving a system of linear inequalities involves individually solving each inequality for y, graphing them on the same axes, and finding the overlapping region, which represents the solution.

5.15 Writing Word Problems for Two-Variable Inequalities

Key Point

The writing process of a word problem involves a real-world scenario, identifying variables, setting up the system of inequalities, and ensuring the problem is solvable.

6. Polynomial Operations and Functions

6.1 Simplifying Polynomial Expressions

🔔 Key Point

When simplifying polynomials, be careful to operate only on terms that have exactly the same variable and power.

🔔 Key Point

The distributive property states that $a(b+c) = ab+ac$. This property is also applied when a polynomial is being multiplied by a monomial.

6.2 Adding and Subtracting Polynomial Expressions

🔔 Key Point

When adding or subtracting polynomials, merge only the like terms, the ones that have the same variable with the same power.

🔔 Key Point

When subtracting polynomials, first distribute the "minus" sign to each of the terms in the polynomial being subtracted, then proceed with addition of the polynomials.

Formula To Remember

1 *Adding Polynomials* 👉 Combine like terms by adding the coefficients: $(ax^n + bx^{n-1} + \ldots) + (cx^n + dx^{n-1} + \ldots) = (a+c)x^n + (b+d)x^{n-1} + \ldots$

2 *Subtracting Polynomials* 👉 Distribute the minus sign and combine like terms: $(ax^n + bx^{n-1} + \ldots) - (cx^n + dx^{n-1} + \ldots) = (a-c)x^n + (b-d)x^{n-1} + \ldots$

6.3 Using Algebra Tiles to Add and Subtract Polynomials

Key Point

Algebra tiles are a visual tool to help you understand adding and subtracting operations in polynomials. The small square is for constant terms, the rectangle for x terms, and the large square for x^2 terms.

6.4 Multiplying Monomial Expressions

Key Point

Multiplying monomials involves multiplying coefficients and adding exponents of similar variables.

6.5 Dividing Monomial Expressions

Key Point

Dividing monomials involves dividing coefficients and subtracting exponents of similar variables.

6.6 Multiplying a Polynomial by a Monomial

Key Point

Multiplying a polynomial by a monomial involves distributing the monomial to each term of the polynomial.

Chapter 6. Polynomial Operations and Functions

6.7 Using Area Models to Multiply Polynomials

Key Point

An area model visually illustrates the process of multiplying each term in one polynomial with each term in the other and then combining the like terms.

6.8 Multiplying Binomial Expressions

Key Point

Multiplication of binomials involves distributing each term in one binomial through each term in the second binomial and combining like terms.

Formula To Remember

FOIL Method for Multiplying Binomials $(a+b)(c+d) = ac + ad + bc + bd$

6.9 Using Algebra Tiles to Multiply Binomials

Key Point

Algebra tiles are rectangular and square shapes that represent variables and constants. A square tile might represent x^2, a rectangular tile could represent x, and a tiny square tile might represent the constant 1.

6.10 Factoring Trinomial Expressions

Key Point

To factor a trinomial $ax^2 + bx + c$, find two numbers p and q such that their product equals $\frac{c}{a}$ and their sum equals $\frac{b}{a}$. The factorized form is $a(x+p)(x+q)$.

Formula To Remember

Factoring Trinomials   For $ax^2 + bx + c$, find p and q such that $pq = \frac{c}{a}$ and $p+q = \frac{b}{a}$. Factorize as $a(x+p)(x+q)$.

6.11 Factoring Polynomial Expressions

Key Point

Factoring a polynomial involves expressing it as a product of two or more simpler polynomial expressions. It requires understanding of the distributive property, greatest common factor, and recognizing patterns in polynomials.

Formula To Remember

1. Difference of Squares ☞ $a^2 - b^2 = (a-b)(a+b)$, where a and b are expressions.
2. Perfect Square Trinomial ☞ $a^2 + 2ab + b^2 = (a+b)^2$ and $a^2 - 2ab + b^2 = (a-b)^2$
3. Sum and Difference of Cubes ☞ $a^3 - b^3 = (a-b)(a^2+ab+b^2)$ and $a^3 + b^3 = (a+b)(a^2-ab+b^2)$

6.12 Using Graphs to Factor Polynomials

Key Point

A factor of a polynomial $p(x)$ is any polynomial $f(x)$ which divides evenly into $p(x)$. If a polynomial $f(x)$ is a factor of $p(x)$, then the root of $f(x)$ will also be the root of $p(x)$. This root will be the x-coordinate at which the graph of $p(x)$ intersects the x-axis.

Key Point

The Factor Theorem states that the polynomial $p(x) = a_n x^n + a_{n-1} x^{n-1} + \ldots + a_1 x + a_0$ has a factor $x - c$ if $p(c) = 0$.

Formula To Remember

1. Factor Theorem ☞ $p(x) = a_n x^n + a_{n-1} x^{n-1} + \ldots + a_1 x + a_0$ has a factor $(x-c)$ if $p(c) = 0$.

6.13 Factoring Special Case Polynomial Expressions

> **Key Point**
>
> Perfect square trinomials are polynomials of the form $a^2 + 2ab + b^2 = (a+b)^2$ or $a^2 - 2ab + b^2 = (a-b)^2$.

> **Key Point**
>
> The 'difference of squares' is a polynomial of the form $a^2 - b^2$, factored as $(a-b)(a+b)$.

6.14 Using Polynomials to Find Perimeter

> **Key Point**
>
> The perimeter of a geometric figure with side lengths given as polynomial expressions can be found by adding all the polynomials.

7. Quadratic Equations and Functions

7.1 Solving Quadratic Equations

Key Point

A quadratic equation $ax^2 + bx + c = 0$ can have either two distinct solutions, one repeated solution, or no real solution, depending on the value of the discriminant $(b^2 - 4ac)$.

Key Point

The discriminant $(b^2 - 4ac)$ determines a quadratic equation's roots: positive, for two distinct real roots, zero, for one repeated real root, negative, for two complex roots. Roots are given by $x = \frac{-b \pm \sqrt{b^2 - 4ac}}{2a}$.

Formula To Remember

1. *Quadratic Formula* ☞ $x = \frac{-b \pm \sqrt{b^2 - 4ac}}{2a}$ where a, b, and c are coefficients of the quadratic equation $ax^2 + bx + c = 0$.

2. *Discriminant* ☞ $\Delta = b^2 - 4ac$ determines the nature of the roots:
   - $\Delta > 0$: Two distinct real roots
   - $\Delta = 0$: Exactly one real root
   - $\Delta < 0$: No real roots (Two complex roots)

7.2 Graphing Quadratic Functions

> **Key Point**
>
> In the quadratic function $y = ax^2 + bx + c$, the parabola always opens upwards when $a > 0$ and downwards when $a < 0$.

> **Key Point**
>
> The highest or the lowest point of a parabola is called the vertex. The vertex of a quadratic function $f(x) = ax^2 + bx + c$ is given by the formula (h,k), where $h = \frac{-b}{2a}$ and $k = f(h)$.

> **Key Point**
>
> The line of symmetry of a parabola passes through the vertex and is represented by the equation $x = h$.

> **Key Point**
>
> The y-intercept of a function is the point where the graph crosses the y-axis. This point is always $(0,c)$, where c is the constant term in the equation of the function.

> **Key Point**
>
> The x-intercept(s) of a function are the points where the graph crosses the x-axis. To find these points, we set $y = 0$ in the equation and solve for x.

Formula To Remember

1. *Standard Form of a Quadratic Function* ☞ $y = f(x) = ax^2 + bx + c$ where a, b, and c are constants.
2. *Vertex of a Parabola* ☞ The vertex (h,k) is given by $h = \frac{-b}{2a}$ and $k = f(h) = a\left(\frac{-b}{2a}\right)^2 + b\left(\frac{-b}{2a}\right) + c$.
3. *Line of Symmetry* ☞ Given by the equation $x = h$ where $h = \frac{-b}{2a}$.
4. *x-Intercepts* ☞ Found by setting $y = 0$ and solving $ax^2 + bx + c = 0$ for x.
5. *y-Intercept* ☞ Found at $(0,c)$ by setting $x = 0$ in the equation.

7.3 Factoring to Solve Quadratic Equations

> **Key Point**
>
> To solve a quadratic equation $x^2 + bx + c = 0$ by factoring, find two numbers that add to b and multiply to c. Use these numbers to factor the equation, then apply the zero product property ($p \times q = 0$ implies $p = 0$ or $q = 0$) to find the solutions.

> **Formula To Remember**
>
> 1. *Standard Form of a Quadratic Equation* ☞ $ax^2 + bx + c = 0$
> 2. *Factoring Quadratics (when $a = 1$)* ☞ Find numbers (p, q) that add to b and multiply to c. Factor to $(x+p)(x+q) = 0$.
> 3. *Zero Product Property* ☞ If $A \times B = 0$, then $A = 0$ or $B = 0$.

7.4 Understanding Transformations of Quadratic Functions

> **Key Point**
>
> If $f(x) = a(x - h)^2 + k$, the coefficient a determines the **vertical stretch or compression** and direction (upwards or downwards) of a parabola. The value of h is related to the **horizontal shift** of the parabola (right or left), and k represents the **vertical shift** (up or down).

Formula To Remember

1 Vertex Form of a Quadratic Function ☞ $f(x) = a(x-h)^2 + k$ where

- a affects the vertical stretch or compression and direction
- h controls the horizontal shift
- k modifies the vertical shift

2 Transformations of a Quadratic Function (Comparison to $f(x) = x^2$) ☞

- If $|a| > 1$, the graph stretches vertically
- If $|a| < 1$, the graph compresses vertically
- If $h > 0$, the graph shifts right
- If $h < 0$, the graph shifts left
- If $k > 0$, the graph shifts up
- If $k < 0$, the graph shifts down

7.5 Completing Function Tables for Quadratic Functions

Key Point

Function tables for quadratic functions are completed by substituting the x values into the quadratic equation to find the corresponding y values.

7.6 Determining Domain and Range of Quadratic Functions

Key Point

The domain of a quadratic functions $f(x) = ax^2 + bx + c$ is all real numbers, $x \in \mathbb{R}$.

Key Point

In the quadratic function $f(x) = ax^2 + bx + c$, if $a > 0$, the range is $[y_{vertex}, +\infty)$. On the other hand, if $a < 0$, the range is $(-\infty, y_{vertex}]$.

Formula To Remember

1. *Domain of a Quadratic Function* 👉 The domain of $f(x) = ax^2 + bx + c$ is all real numbers, $x \in \mathbb{R}$.
2. *Range of a Quadratic Function (opens upwards)* 👉 If $a > 0$, the range is $[y_{vertex}, +\infty)$.
3. *Range of a Quadratic Function (opens downwards)* 👉 If $a < 0$, the range is $(-\infty, y_{vertex}]$.

7.7 Using Algebra Tiles to Factor Quadratics
Key Point

Algebra tiles visually represent algebraic expressions. For quadratics, x^2, x, and 1 are respectively represented by square, rectangular, and small square tiles.

7.8 Writing Quadratic Functions from Vertices and Points
Key Point

$f(x) = a(x-h)^2 + k$ is the vertex form of a quadratic function where (h,k) is the vertex and a is a non-zero constant. The vertex form can be written using given vertex and a point on the function.

8. Relations and Functions

8.1 Understanding Function Notation and Evaluation

Key Point

Functions are mathematical operations that assign unique outputs to given inputs. Here, instead of using y, we use $f(x)$, where x is the input and $f(x)$ is the output of the function.

Key Point

When we evaluate functions, we substitute specific values of x into the function formula to find the corresponding output, $f(x)$.

8.2 Completing a Function Table from an Equation

Key Point

A function table helps us understand the behavior of a function at different input (x) values by listing down the corresponding output (y) values.

8.3 Determining Domain and Range of Relations

Key Point

The domain of a relation are the x-values (or inputs), while the range are the y-values (or outputs).

8.4 Performing Addition and Subtraction of Functions

Key Point

The addition $(f+g)(x)$ or subtraction $(f-g)(x)$ of two functions gives us a new function where addition or subtraction has been performed at each value of x.

Formula To Remember

1. Addition of Functions 👉 $(f+g)(x) = f(x) + g(x)$
2. Subtraction of Functions 👉 $(f-g)(x) = f(x) - g(x)$

8.5 Performing Multiplication and Division of Functions

Key Point

Multiplication $(f \times g)(x)$ or division $(\frac{f}{g})(x)$ (for $g(x) \neq 0$) of two functions gives us a new function where multiplication or division has been performed at each functional value.

Formula To Remember

1. Multiplication of Functions 👉 $(f \times g)(x) = f(x) \times g(x)$
2. Division of Functions 👉 $(\frac{f}{g})(x) = \frac{f(x)}{g(x)}$, provided $g(x) \neq 0$

8.6 Composing Functions

Key Point

Function composition $(f \circ g)(x)$ is a process where the output of one function, $g(x)$, is used as the input for another function, $f(x)$.

Formula To Remember

1. Function Composition 👉 $(f \circ g)(x) = f(g(x))$

8.7 Evaluating Exponential Functions

🔔 Key Point

Calculating the value of an exponential function involves substituting the input into the exponent of the base.

📌 Formula To Remember

1. *General Form of an Exponential Function* 👉 $f(x) = a \cdot b^x$ where $a \neq 0$ and $b > 0, b \neq 1$.
2. *y-Intercept of Exponential Function* 👉 The *y*-intercept is at $(0, a)$.

8.8 Matching Exponential Functions with Graphs

🔔 Key Point

An exponential function $f(x) = a \cdot b^x$ displays a 'J' shaped graph. It exhibits exponential growth for $b > 1$ and decay for $0 < b < 1$. The *y*-intercept of the graph is at a.

8.9 Writing Exponential Functions from Word Problems

🔔 Key Point

The initial value in a word problem usually signifies the *y*-intercept 'a'. The rate of growth or decay identified in the problem is the base 'b'.

📌 Formula To Remember

1. *General Exponential Function* 👉 $f(x) = a \cdot b^x$ where a is the initial value, and b is the rate of growth or decay.

8.10 Understanding Function Inverses

🔔 Key Point

To find the inverse of $f(x)$, replace $f(x)$ with y, then interchange x and y and solve the resulting equation for y. Remember not every function has an inverse. At least, it should be a one-to-one function which means that every element in the domain is mapped to a unique element in the range.

📌 Formula To Remember

1 *Inverse of a Function* 👉 If $y = f(x)$, then $x = f^{-1}(y)$. Swap x and y and solve for y to find $f^{-1}(x)$.

8.11 Understanding Rate of Change and Slope

🔔 Key Point

For every function we have:

$$\text{Rate of Change} = \frac{\text{Change in Output}(Y)}{\text{Change in Input}(X)} = \frac{\Delta Y}{\Delta X}.$$

For linear functions, rate of change is the slope of the line.

📌 Formula To Remember

1 *Rate of Change Formula* 👉 Rate of Change $= \frac{\Delta Y}{\Delta X}$ where ΔY is the change in output and ΔX is the change in input.

2 *Slope Formula* 👉 Slope $= \frac{y_2 - y_1}{x_2 - x_1}$ where (x_1, y_1) and (x_2, y_2) are points on the line.

9. Radical Expressions and Equations

9.1 Simplifying Radical Expressions

🔔 Key Point

To simplify a radical expression, find the prime factors of the radicand and use radical properties:
$\sqrt[n]{x^a} = x^{\frac{a}{n}}$, $\quad \sqrt[n]{xy} = \sqrt[n]{x} \times \sqrt[n]{y}$, $\quad \sqrt[n]{\frac{x}{y}} = \frac{\sqrt[n]{x}}{\sqrt[n]{y}}$.

📌 Formula To Remember

1 Basic Radical Properties 👉 $\sqrt[n]{x^a} = x^{\frac{a}{n}}$, $\sqrt[n]{xy} = \sqrt[n]{x} \times \sqrt[n]{y}$, $\sqrt[n]{\frac{x}{y}} = \frac{\sqrt[n]{x}}{\sqrt[n]{y}}$

9.2 Performing Addition and Subtraction of Radical Expressions

🔔 Key Point

Like radicals are radical expressions that have the same radicand and index. Addition and subtraction operations are only valid between like radicals.

Formula To Remember

1. *Simplifying Like Radicals* 👉 $a\sqrt[n]{b} + c\sqrt[n]{b} = (a+c)\sqrt[n]{b}$ where n is the index, b is the radicand, and a, c are coefficients.

2. *Simplifying Unlike Radicals* 👉 Expressions like $a\sqrt[n]{b} + c\sqrt[n]{d}$ cannot be simplified if $b \neq d$ even if n is the same.

9.3 Performing Multiplication of Radical Expressions

Key Point

The product rule for radicals states:

Given $a, b \geq 0$ and n is a positive even integer, then $\sqrt[n]{a}\sqrt[n]{b} = \sqrt[n]{ab}$.

Given a, b are real numbers and n is a positive odd integer, then $\sqrt[n]{a}\sqrt[n]{b} = \sqrt[n]{ab}$.

Formula To Remember

1. *Product Rule for Radicals* 👉 $\sqrt[n]{a}\sqrt[n]{b} = \sqrt[n]{ab}$

9.4 Rationalizing Radical Expressions

Key Point

Rationalizing a denominator involves multiplying it by a radical that will eliminate the square root or cube root, leaving only rational numbers. For binomial denominators, we employ the conjugate to eliminate the square roots in the denominator.

Formula To Remember

1. *Rationalizing the Denominator* 👉 For a simple radical: $\frac{a}{\sqrt{b}}$, multiply by $\frac{\sqrt{b}}{\sqrt{b}}$ to get $\frac{a\sqrt{b}}{b}$.

2. *Rationalizing Binomial Denominators* 👉 For a binomial $\frac{a}{\sqrt{b}-c}$, multiply by the conjugate $\frac{\sqrt{b}+c}{\sqrt{b}+c}$ to get $\frac{a(\sqrt{b}+c)}{b-c^2}$.

9.5 Solving Radical Equations

> **Key Point**
>
> To solve radical equations, isolate the radical on one side. Then square or raise to index for both sides of the equation to eliminate the radical. Solve the equation. Always check solutions to avoid extraneous solutions.

9.6 Determining Domain and Range of Radical Functions

> **Key Point**
>
> The domain of a radical function with an even root is only those x values that result in a number greater than or equal to zero under the radical. For an odd root, the domain is equal to the domain of the expression under the radical.

> **Key Point**
>
> The range of a radical function depends on whether the root is even or odd. For a radical function of the form $y = c\sqrt[n]{ax+b} + k$, where n is an even integer, the range depends on the sign of c. If $c > 0$, the range is $y \geq k$. If $c < 0$, the range is $y \leq k$, assuming $ax+b$ is non-negative to ensure the radical is defined. For odd indices, the range is all real numbers, reflecting the unbounded nature of odd root functions over the domain of all real numbers.

Formula To Remember

1. *Domain of Even Root Radical Functions* ☞ For $y = c\sqrt[n]{ax+b} + k$ with n even, domain: $ax+b \geq 0$.
2. *Range of Even Root Radical Functions* ☞ For $y = c\sqrt[n]{ax+b} + k$ with n even, if $c > 0$, then range: $y \geq k$. If $c < 0$, then range: $y \leq k$.
3. *Domain and Range of Odd Root Radical Functions* ☞ For $y = c\sqrt[n]{ax+b} + k$ with n odd, domain: All real numbers ($\mathbb{R}$). Range: All real numbers ($\mathbb{R}$).

9.7 Simplifying Radicals with Fractional Components

Key Point

To simplify a radical fraction, factorize the numerator and denominator, and apply $\sqrt[n]{\frac{a}{b}} = \frac{\sqrt[n]{a}}{\sqrt[n]{b}}$, where a and b are nonnegative if n is even. Simplify each radical by extracting perfect powers.

Formula To Remember

1. *Simplification of Fractional Radicals* 👉 $\sqrt[n]{\frac{a}{b}} = \frac{\sqrt[n]{a}}{\sqrt[n]{b}}$ where $b \neq 0$.

10. Rational Expressions

10.1 Simplifying Complex Fractions

🔔 Key Point

To simplify a complex fraction, unify denominators in numerator and denominator by finding the LCD. Simplify numerator and denominator of the complex fraction separately using this LCD. Carry out the division operation in the complex fraction with simplified numerator and denominator.

10.2 Graphing Rational Functions

🔔 Key Point

The domain of a rational function consists of all real numbers except for the zeroes of the denominator $q(x)$.

🔔 Key Point

A vertical asymptote is a vertical line $x = a$, where a is a root of the denominator $q(x)$. The function tends towards $\pm\infty$ when x approaches a.

🔔 Key Point

A horizontal asymptote is a horizontal line $y = b$, where $b = \frac{p(n)}{q(n)}$ for large n if the degree of $p(x)$ is less than or equal to the degree of $q(x)$.

10.3 Performing Addition and Subtraction of Rational Expressions

🔔 Key Point

The x-intercepts are the roots of the numerator $p(x)$ and the y-intercept is found by substituting $x = 0$ into the function.

📌 Formula To Remember

1. *Rational Function Form* 👉 $f(x) = \frac{p(x)}{q(x)}$ where $p(x)$ and $q(x)$ are polynomials and $q(x) \neq 0$.
2. *Domain of a Rational Function* 👉 The domain is all real numbers except the roots of $q(x)$.
3. *Vertical Asymptote* 👉 Vertical line $x = a$ where a is a root of $q(x)$.
4. *Horizontal Asymptote* 👉 Horizontal line $y = b$ where $b = \frac{p(n)}{q(n)}$ for large n ($n \to \pm\infty$), if degree of $p(x) \leq$ degree of $q(x)$.

10.3 Performing Addition and Subtraction of Rational Expressions

🔔 Key Point

The Least Common Denominator (LCD) of rational expressions is found by determining the Least Common Multiple (LCM) of the polynomials in the denominator.

🔔 Key Point

Before adding or subtracting, rational expressions need to have a common denominator. Each expression is rewritten with the LCD as the denominator. Then, add or subtract the numerators and simplify as needed.

10.4 Performing Multiplication of Rational Expressions

🔔 Key Point

The process of multiplying rational expressions involves factoring, identifying common factors in the numerators and denominators, and cancelling them out.

10.5 Performing Division of Rational Expressions

> **Key Point**
>
> The division of rational expressions requires factorization of expressions, changing division to multiplication by flipping the divisor, and cancelling out common factors.

10.6 Evaluating Integers with Rational Exponents

> **Key Point**
>
> An integer with a rational exponent $a^{\frac{m}{n}}$ can be evaluated by raising a to the power m and taking the nth root of the result, or by taking the nth root of a and raising the result to the power m.

11. Number Sequences and Formulas

11.1 Evaluating Variable Expressions for Number Sequences

Key Point

Variable expressions hold a general form such as $A_n = 2n - 1$, where A_n represents the n-th term of the sequence and n is the position of each term.

Formula To Remember

1. *General Form of a Sequence* $A_n = f(n)$ where A_n is the n-th term and $f(n)$ is a function representing the formula of the sequence.

11.2 Writing Variable Expressions for Arithmetic Sequences

Key Point

The variable expression for an arithmetic sequence is $A_n = a + (n-1)d$ where a is the first term, d is the common difference, and n is the position of the term.

Chapter 11. Number Sequences and Formulas

Formula To Remember

General Form of an Arithmetic Sequence ☞ $A_n = a + (n-1)d$ where a is the first term, d is the common difference, and n is the term number.

11.3 Writing Variable Expressions for Geometric Sequences

Key Point

The variable expression for a geometric sequence is $G_n = a \times r^{n-1}$, where a is the first term, r is the common ratio and n is the position of the term.

Formula To Remember

Expression for the n-th Term of a Geometric Sequence ☞ $G_n = a \times r^{n-1}$ where a is the first term, r is the common ratio, and n is the term number.

11.4 Evaluating Recursive Formulas for Sequences

Key Point

A recursive formula defines each term of a sequence using its predecessors. Evaluation starts from known initial term(s) and proceeds via the recursive relation.

11.5 Formulating a Recursive Sequence

Key Point

To derive a recursive formula, identify initial terms, observe patterns in differences or ratios between terms, and formulate these observations into a function describing each term by its predecessors.

11.5 Formulating a Recursive Sequence

 Formula To Remember

1. *Recursive Formula for Arithmetic Sequence* ☞ $a_n = a_{n-1} + d$ where d is the common difference.

2. *Recursive Formula for Geometric Sequence* ☞ $a_n = a_{n-1} \times r$ where r is the common ratio.

3. *Recursive Formula for Fibonacci Sequence* ☞ $a_n = a_{n-1} + a_{n-2}$ for $n \geq 3$, with $a_1 = a_2 = 1$.

12. Direct and Inverse Variations

12.1 Determining the Constant of Variation

🔔 Key Point

The constant of variation in a direct variation is determined by the equation $y = kx$, where k is calculated by the division $\frac{y}{x}$. For an inverse variation, it is given by $xy = k$, where k can be found by xy.

📌 Formula To Remember

1. *Direct Variation Formula* 👉 $y = kx$ where $k = \frac{y}{x}$ and k is the constant of variation.
2. *Inverse Variation Formula* 👉 $xy = k$ where k can be found using $k = xy$ and k is the constant of variation.

12.2 Formulating and Solving Direct Variation Equations

🔔 Key Point

In a direct variation, the value of one variable changes directly with the other. A change in one variable results in a proportional change in the other variable.

12.3 Modeling Inverse Variation

> **Key Point**
>
> The product of the variables x and y in an inverse variation is always constant; $xy = k$. The greater the value of x, the smaller the value of y, and vice versa.

13. Statistical Analysis and Probability

13.1 Calculating Mean, Median, Mode, and Range

Key Point

The **mean** is the sum of data divided by the total number of data. The **median** is the middle value in ordered data. The **mode** is the most frequent value(s). The **range** is the difference between the highest and lowest values.

Formula To Remember

1. *Mean (Average)* ☞ Mean $= \frac{x_1 + x_2 + \cdots + x_n}{n}$
2. *Median* ☞ Order data. For odd n, median = middle value. For even n, median = average of two middle values.
3. *Mode* ☞ Most frequently occurring value(s) in the data set.
4. *Range* ☞ Range = Maximum − Minimum

13.2 Creating a Pie Graph

Key Point

A pie graph is used to visually represent percentages or proportional data and are divided into sectors, where each sector represents a proportion of the total.

13.3 Analyzing Scatter Plots

> **Key Point**
>
> To create a pie graph, calculate the total of all data, then determine each slice's angle using $\frac{\text{data value}}{\text{total data sum}} \times 360°$. Draw a circle, plot each slice by its angle, and label accordingly.

> **Formula To Remember**
>
> 1. *Calculating the Angle of a Pie Slice* 👉 $\text{angle} = \frac{\text{data value}}{\text{total data sum}} \times 360°$

13.3 Analyzing Scatter Plots

> **Key Point**
>
> In scatter plots, a positive trend occurs when y increases with x, and a negative trend occurs when y decreases as x increases.

13.4 Calculating and Interpreting Correlation Coefficients

> **Key Point**
>
> A correlation coefficient quantifies the degree to which two variables are related. A value of $+1$ implies a perfect positive correlation, while -1 implies a perfect negative correlation. A value of 0 implies no correlation.

> **Key Point**
>
> The sample Pearson correlation coefficient (r) measures the linear relationship between two variables, calculated by
>
> $$r = \frac{1}{n-1} \sum_{i=1}^{n} \left(\frac{x_i - \bar{x}}{s_x}\right)\left(\frac{y_i - \bar{y}}{s_y}\right).$$
>
> Where x and y are the variables we are considering, $\bar{x}$ and $\bar{y}$ are their means, and s_x and s_y are their sample standard deviations.

> **Formula To Remember**
>
> 1. *Sample Standard Deviation* 👉 $s_x = \sqrt{\frac{1}{n-1} \sum_{i=1}^{n} (x_i - \bar{x})^2}$
> 2. *Sample Pearson Correlation Coefficient* 👉 $r = \frac{1}{n-1} \sum_{i=1}^{n} \left(\frac{x_i - \bar{x}}{s_x}\right)\left(\frac{y_i - \bar{y}}{s_y}\right)$

13.5 Formulating the Equation of a Regression Line

Key Point

A regression line minimizes the sum of squares of vertical distances from each data point to the line. Its equation, $y = mx + b$, where $m = r \times \frac{s_y}{s_x}$ and $b = \bar{y} - m\bar{x}$, depends on the correlation coefficient r, standard deviations s_x, s_y, and means $\bar{x}, \bar{y}$.

Formula To Remember

1. Regression Line Equation $y = mx + b$ where $m = r \times \frac{s_y}{s_x}$ and $b = \bar{y} - m\bar{x}$.

13.6 Understanding Correlation and Causation

Key Point

Correlation does not imply causation. A statistical relationship between variables, known as correlation, does not mean that one variable is the cause of changes in the other.

13.7 Understanding Permutations and Combinations

Key Point

In permutations, the order of selection is important whereas in combinations, order does not matter. The formula for permutations $P(n,r) = \frac{n!}{(n-r)!}$ and for combinations $C(n,r) = \frac{n!}{r!(n-r)!}$, where $n!$ denotes the factorial of n.

Formula To Remember

1. Permutations Formula $P(n,r) = \frac{n!}{(n-r)!}$, where $n!$ denotes the factorial of n.
2. Combinations Formula $C(n,r) = \frac{n!}{r!(n-r)!}$

13.8 Solving Probability Problems

Key Point

The probability of event A:

$$P(A) = \frac{\text{number of desired outcomes}}{\text{total number of possible outcomes}}.$$

Key Point

When we multiply the probabilities of two independent events, we get the probability that both occur. This can be represented by $P(A \text{ and } B) = P(A) \times P(B)$, where A and B are independent events.

Formula To Remember

1. *Basic Probability Formula* $P(A) = \frac{\text{number of desired outcomes}}{\text{total number of possible outcomes}}$
2. *Probability of Independent Events* $P(A \text{ and } B) = P(A) \times P(B)$ where A and B are independent events.

14. Formula Sheet for All Topics

For your benefit we are providing all formulas again in one place. This is a quick reference guide to the formulas you need to remember for Algebra topics. We recommend that you review this formula sheet before taking the practice tests. This will help you recall and apply the formulas effectively.

Formulas For Chapter: Foundations and Basic Principles

Adding and Subtracting Integers

1. *Adding Two Negative Integers* ☞ $(-a)+(-b) = -(a+b)$ (both a and b are positive).
2. *Adding Positive and Negative Integers* ☞ $a+(-b)$ is equal to $a-b$ if $a>b$, and is equal to $-(b-a)$ if $b>a$, where a and b are positive.
3. *Subtracting Integers Using Addition* ☞ $a-b = a+(-b)$

Multiplying and Dividing Integers

1. *Rules for Multiplying Integers* ☞ Let a and b be integers.

   $a \times b$ is positive if a and b have the same sign.

   $a \times b$ is negative if a and b have opposite signs.

2. *Rules for Dividing Integers* ☞ Let a and b be integers, $b \neq 0$.

   $\frac{a}{b}$ is positive if a and b have the same sign.

   $\frac{a}{b}$ is negative if a and b have opposite signs.

Order of Operations

1 *PEMDAS Rule* 👉 Perform operations in this order: 1. Parentheses, 2. Exponents, 3. Multiplication and Division (from left to right), 4. Addition and Subtraction (from left to right).

The Distributive Property

1 *Distributive Property* 👉 $a \times (b+c) = a \times b + a \times c$

2 *Distributive Property with Subtraction* 👉 $a \times (b-c) = a \times b - a \times c$

Integers and Absolute Value

1 *Definition of Absolute Value* 👉 $|a| = \begin{cases} a & \text{if } a \geq 0 \\ -a & \text{if } a < 0. \end{cases}$

Proportional Ratios

1 *Cross-Multiplication* 👉 If $\frac{a}{b} = \frac{c}{d}$, then $ad = bc$.

Similarity and Ratios

1 *Scale Factor in Similar Figures* 👉 If two figures are similar with scale factor k, then the ratio of any corresponding linear measurements (sides, perimeters, etc.) is k.

2 *Area Ratio in Similar Figures* 👉 If two figures are similar with scale factor k, the ratio of their areas is k^2.

3 *Volume Ratio in Similar Figures* 👉 If two figures are similar with scale factor k, the ratio of their volumes is k^3.

Percent Problems

1 *Base Formula for Percent Problems* 👉 $b \times \frac{r}{100} = p$ where b is the base, r is the percent rate, and p is the percentage amount.

2 *Solving for the Base* 👉 $b = \frac{p}{\frac{r}{100}}$ or equivalently $b = \frac{p \times 100}{r}$

Chapter 14. Formula Sheet for All Topics

Percent of Increase and Decrease

1) *Percent Increase* ☞ Percent Increase $= \frac{\text{Final Value} - \text{Initial Value}}{\text{Initial Value}} \times 100\%$

2) *Percent Decrease* ☞ Percent Decrease $= \frac{\text{Initial Value} - \text{Final Value}}{\text{Initial Value}} \times 100\%$

Discount, Tax, and Tip

1) *Discount Calculation* ☞ Discount $= \frac{\text{Discount Rate}}{100} \times \text{Mark Price}$

2) *Tax Calculation* ☞ Tax $= \frac{\text{Tax Rate}}{100} \times \text{Price}$

3) *Tip Calculation* ☞ Tip $= \frac{\text{Tip Rate}}{100} \times \text{Total Bill}$

Simple Interest

1) *Simple Interest Formula* ☞ $I = P \times R \times T$ where I is the interest, P the principal, R the rate of interest, and T the time in years.

Formulas For Chapter: Exponents and Variables

Mastering Exponent Multiplication

1) *Product of Powers Property* ☞ $a^n \times a^m = a^{n+m}$

Exploring Powers of Products

1) *Power of a Product Property* ☞ $(ab)^n = a^n b^n$

Mastering Exponent Division

1) *Quotient of Powers Property* ☞ $b^m \div b^n = b^{m-n}$ where $b \neq 0$.

Understanding Zero and Negative Exponents

1. *Zero Exponent Rule* 👉 $a^0 = 1$ where a is any non-zero number.
2. *Negative Exponent Rule* 👉 $a^{-n} = \frac{1}{a^n}$ where a is any non-zero number and n is a positive integer.

Working with Negative Bases

1. *Power Rule for Negative Bases* 👉 $(-a)^n = \begin{cases} a^n & \text{if } n \text{ is even} \\ -a^n & \text{if } n \text{ is odd} \end{cases}$

Introduction to Scientific Notation

1. *Scientific Notation Form* 👉 $a \times 10^n$ where $1 \leq |a| < 10$ and n is an integer.

Multiplication and Division in Scientific Notation

1. *Multiplying in Scientific Notation* 👉 $(a \times 10^b) \times (c \times 10^d) = (a \times c) \times 10^{b+d}$
2. *Dividing in Scientific Notation* 👉 $\frac{a \times 10^b}{c \times 10^d} = \left(\frac{a}{c}\right) \times 10^{b-d}$

Formulas For Chapter: Algebraic Expressions and Equations

Translate a Phrase into an Algebraic Statement

1. *Algebraic Representation of Addition* 👉 Phrase "the sum of a and b" translates to $a + b$.
2. *Algebraic Representation of Multiplication* 👉 Phrase "product of a and b" translates to $a \times b$.
3. *Algebraic Representation of Division* 👉 Phrase "quotient of a and b" translates to $\frac{a}{b}$.
4. *Algebraic Representation of Equality* 👉 Phrase "a is b" translates to $a = b$.

Simplifying Variable Expressions

1. *Order of Operations - PEMDAS* 👉 Parenthesis, Exponents, Multiplication & Division (left to right), Addition & Subtraction (left to right).

2. *Distributive Property* 👉 $a(b+c) = ab+ac$

Solving One-Step Equations

1. *Solving One-Step Addition or Subtraction Equations* 👉 For $x+a=b$, solve by subtracting a from both sides: $x = b-a$.

2. *Solving One-Step Multiplication or Division Equations* 👉 For $ax=b$, solve by dividing both sides by a: $x = \frac{b}{a}$.

Finding Midpoint

1. *Midpoint Formula* 👉 $M = \left(\frac{x_1+x_2}{2}, \frac{y_1+y_2}{2}\right)$ where (x_1, y_1) and (x_2, y_2) are the coordinates of the endpoints.

Finding Distance of Two Points

1. *Distance Formula* 👉 $d = \sqrt{(x_2-x_1)^2 + (y_2-y_1)^2}$

Formulas For Chapter — Introduction to Linear Functions

Determining Slopes

1. *Slope Formula* 👉 $m = \frac{y_2-y_1}{x_2-x_1}$ where (x_1, y_1) and (x_2, y_2) are coordinates of two points on the line.

Formulating Linear Equations

1. *Slope-Intercept Form of a Line* 👉 $y = mx+b$ where m is the slope and b is the y-intercept.

2. *Finding the y-Intercept from a Point* 👉 $b = y_1 - mx_1$ where (x_1, y_1) is a point on the line and m is the slope.

Understanding Slope-Intercept and Point-Slope Forms

☞ *Point-Slope Form of a Line* 👉 $y - y_1 = m(x - x_1)$ where m is the slope and (x_1, y_1) is a known point on the line.

Identifying x- and y-intercepts

☞ *Finding the x-intercept* 👉 Set $y = 0$ in the equation of the line and solve for x.

☞ *Finding the y-intercept* 👉 Set $x = 0$ in the equation of the line and solve for y.

Graphing Standard Form Equations

☞ *Standard Form of a Linear Equation* 👉 $Ax + By = C$ where A, B, and C are constants, with A and B not both zero.

Understanding Horizontal and Vertical Lines

☞ *Equation of a Horizontal Line* 👉 $y = b$ where b is the y-coordinate of all points on the line.

☞ *Equation of a Vertical Line* 👉 $x = a$ where a is the x-coordinate of all points on the line.

Understanding Parallel and Perpendicular Lines

☞ *Condition for Parallel Lines* 👉 Two lines are parallel if their slopes are equal, represented as $m_1 = m_2$, and they do not have the same y-intercept.

☞ *Condition for Perpendicular Lines* 👉 Two lines are perpendicular if the slope of one line is the negative reciprocal of the other, expressed as $m_2 = -\frac{1}{m_1}$.

Graphing Absolute Value Equations

☞ *Standard Form of Absolute Value Equation* 👉 $y = |x|$ produces a "V" shaped graph with vertex at $(0,0)$.

☞ *Effect of Horizontal Shift on Vertex* 👉 $y = |x - h|$ moves the vertex to $(h, 0)$.

Formulas for Chapter: Inequalities and Systems of Equations

Solving One-Step Inequalities

1. *Solving Addition or Subtraction Inequalities* ☞ $x+c > b$ becomes $x > b-c$; $x-c < b$ becomes $x < b+c$.

2. *Solving Multiplication or Division Inequalities* ☞ $ax < b$ becomes $x < \frac{b}{a}$ and flip the inequality if $a < 0$; $ax > b$ becomes $x > \frac{b}{a}$ and flip the inequality if $a < 0$.

Writing Linear Inequalities from Graphs

1. *Form of a Linear Inequality* ☞ $y \leq mx+b$ or $y \geq mx+b$ for a solid line, $y < mx+b$ or $y > mx+b$ for a dashed line.

Solving Absolute Value Inequalities

1. *"Greater than" Absolute Value Inequality* ☞ $|A| > B$ is solved by $A > B$ or $A < -B$.

2. *"Less than" Absolute Value Inequality* ☞ $|A| < B$ is solved by $-B < A < B$.

Determining the Number of Solutions to Linear Equations

1. *Criteria for Unique Solution* ☞ A system has exactly one solution if $\frac{a_1}{a_2} \neq \frac{b_1}{b_2}$, considering equations of form $a_1x + b_1y = c_1$ and $a_2x + b_2y = c_2$.

2. *Criteria for No Solution* ☞ A system has no solution if $\frac{a_1}{a_2} = \frac{b_1}{b_2} \neq \frac{c_1}{c_2}$, where equations are in the form $a_1x + b_1y = c_1$ and $a_2x + b_2y = c_2$.

3. *Criteria for Infinitely Many Solutions* ☞ A system has infinitely many solutions if $\frac{a_1}{a_2} = \frac{b_1}{b_2} = \frac{c_1}{c_2}$ for equations $a_1x + b_1y = c_1$ and $a_2x + b_2y = c_2$.

Formulas for Chapter: Polynomial Operations and Functions

Adding and Subtracting Polynomial Expressions

1 *Adding Polynomials* 👉 Combine like terms by adding the coefficients: $(ax^n + bx^{n-1} + \ldots) + (cx^n + dx^{n-1} + \ldots) = (a+c)x^n + (b+d)x^{n-1} + \ldots$

2 *Subtracting Polynomials* 👉 Distribute the minus sign and combine like terms: $(ax^n + bx^{n-1} + \ldots) - (cx^n + dx^{n-1} + \ldots) = (a-c)x^n + (b-d)x^{n-1} + \ldots$

Multiplying Binomial Expressions

1 *FOIL Method for Multiplying Binomials* 👉 $(a+b)(c+d) = ac + ad + bc + bd$

Factoring Trinomial Expressions

1 *Factoring Trinomials* 👉 For $ax^2 + bx + c$, find p and q such that $pq = \frac{c}{a}$ and $p + q = \frac{b}{a}$. Factorize as $a(x+p)(x+q)$.

Factoring Polynomial Expressions

1 *Difference of Squares* 👉 $a^2 - b^2 = (a-b)(a+b)$, where a and b are expressions.

2 *Perfect Square Trinomial* 👉 $a^2 + 2ab + b^2 = (a+b)^2$ and $a^2 - 2ab + b^2 = (a-b)^2$

3 *Sum and Difference of Cubes* 👉 $a^3 - b^3 = (a-b)(a^2 + ab + b^2)$ and $a^3 + b^3 = (a+b)(a^2 - ab + b^2)$

Using Graphs to Factor Polynomials

1 *Factor Theorem* 👉 $p(x) = a_n x^n + a_{n-1} x^{n-1} + \ldots + a_1 x + a_0$ has a factor $(x-c)$ if $p(c) = 0$.

Formulas for Chapter: Quadratic Equations and Functions

Chapter 14. Formula Sheet for All Topics

Solving Quadratic Equations

1. *Quadratic Formula* ☞ $x = \frac{-b \pm \sqrt{b^2 - 4ac}}{2a}$ where a, b, and c are coefficients of the quadratic equation $ax^2 + bx + c = 0$.

2. *Discriminant* ☞ $\Delta = b^2 - 4ac$ determines the nature of the roots:
   - $\Delta > 0$: Two distinct real roots
   - $\Delta = 0$: Exactly one real root
   - $\Delta < 0$: No real roots (Two complex roots)

Graphing Quadratic Functions

1. *Standard Form of a Quadratic Function* ☞ $y = f(x) = ax^2 + bx + c$ where a, b, and c are constants.
2. *Vertex of a Parabola* ☞ The vertex (h, k) is given by $h = \frac{-b}{2a}$ and $k = f(h) = a\left(\frac{-b}{2a}\right)^2 + b\left(\frac{-b}{2a}\right) + c$.
3. *Line of Symmetry* ☞ Given by the equation $x = h$ where $h = \frac{-b}{2a}$.
4. *x-Intercepts* ☞ Found by setting $y = 0$ and solving $ax^2 + bx + c = 0$ for x.
5. *y-Intercept* ☞ Found at $(0, c)$ by setting $x = 0$ in the equation.

Factoring to Solve Quadratic Equations

1. *Standard Form of a Quadratic Equation* ☞ $ax^2 + bx + c = 0$
2. *Factoring Quadratics (when $a = 1$)* ☞ Find numbers (p, q) that add to b and multiply to c. Factor to $(x + p)(x + q) = 0$.
3. *Zero Product Property* ☞ If $A \times B = 0$, then $A = 0$ or $B = 0$.

Understanding Transformations of Quadratic Functions

1 *Vertex Form of a Quadratic Function* 👉 $f(x) = a(x-h)^2 + k$ where
- a affects the vertical stretch or compression and direction
- h controls the horizontal shift
- k modifies the vertical shift

2 *Transformations of a Quadratic Function (Comparison to $f(x) = x^2$)* 👉
- If $|a| > 1$, the graph stretches vertically
- If $|a| < 1$, the graph compresses vertically
- If $h > 0$, the graph shifts right
- If $h < 0$, the graph shifts left
- If $k > 0$, the graph shifts up
- If $k < 0$, the graph shifts down

Determining Domain and Range of Quadratic Functions

1 *Domain of a Quadratic Function* 👉 The domain of $f(x) = ax^2 + bx + c$ is all real numbers, $x \in \mathbb{R}$.

2 *Range of a Quadratic Function (opens upwards)* 👉 If $a > 0$, the range is $[y_{vertex}, +\infty)$.

3 *Range of a Quadratic Function (opens downwards)* 👉 If $a < 0$, the range is $(-\infty, y_{vertex}]$.

📖 Formulas For Chapter: Relations and Functions

Performing Addition and Subtraction of Functions

1 *Addition of Functions* 👉 $(f+g)(x) = f(x) + g(x)$

2 *Subtraction of Functions* 👉 $(f-g)(x) = f(x) - g(x)$

Performing Multiplication and Division of Functions

1 *Multiplication of Functions* 👉 $(f \times g)(x) = f(x) \times g(x)$

2 *Division of Functions* 👉 $\left(\frac{f}{g}\right)(x) = \frac{f(x)}{g(x)}$, provided $g(x) \neq 0$

Composing Functions

1 *Function Composition* 👉 $(f \circ g)(x) = f(g(x))$

Evaluating Exponential Functions

1 *General Form of an Exponential Function* 👉 $f(x) = a \cdot b^x$ where $a \neq 0$ and $b > 0, b \neq 1$.

2 *y-Intercept of Exponential Function* 👉 The y-intercept is at $(0, a)$.

Writing Exponential Functions from Word Problems

1 *General Exponential Function* 👉 $f(x) = a \cdot b^x$ where a is the initial value, and b is the rate of growth or decay.

Understanding Function Inverses

1 *Inverse of a Function* 👉 If $y = f(x)$, then $x = f^{-1}(y)$. Swap x and y and solve for y to find $f^{-1}(x)$.

Understanding Rate of Change and Slope

1 *Rate of Change Formula* 👉 Rate of Change $= \frac{\Delta Y}{\Delta X}$ where ΔY is the change in output and ΔX is the change in input.

2 *Slope Formula* 👉 Slope $= \frac{y_2 - y_1}{x_2 - x_1}$ where (x_1, y_1) and (x_2, y_2) are points on the line.

Formulas For Chapter: Radical Expressions and Equations

Simplifying Radical Expressions

1 *Basic Radical Properties* 👉 $\sqrt[n]{x^a} = x^{\frac{a}{n}}$, $\sqrt[n]{xy} = \sqrt[n]{x} \times \sqrt[n]{y}$, $\sqrt[n]{\frac{x}{y}} = \frac{\sqrt[n]{x}}{\sqrt[n]{y}}$

Performing Addition and Subtraction of Radical Expressions

1. *Simplifying Like Radicals* ☞ $a\sqrt[n]{b} + c\sqrt[n]{b} = (a+c)\sqrt[n]{b}$ where n is the index, b is the radicand, and a, c are coefficients.

2. *Simplifying Unlike Radicals* ☞ Expressions like $a\sqrt[n]{b} + c\sqrt[n]{d}$ cannot be simplified if $b \neq d$ even if n is the same.

Performing Multiplication of Radical Expressions

1. *Product Rule for Radicals* ☞ $\sqrt[n]{a}\sqrt[n]{b} = \sqrt[n]{ab}$

Rationalizing Radical Expressions

1. *Rationalizing the Denominator* ☞ For a simple radical: $\frac{a}{\sqrt{b}}$, multiply by $\frac{\sqrt{b}}{\sqrt{b}}$ to get $\frac{a\sqrt{b}}{b}$.

2. *Rationalizing Binomial Denominators* ☞ For a binomial $\frac{a}{\sqrt{b}-c}$, multiply by the conjugate $\frac{\sqrt{b}+c}{\sqrt{b}+c}$ to get $\frac{a(\sqrt{b}+c)}{b-c^2}$.

Determining Domain and Range of Radical Functions

1. *Domain of Even Root Radical Functions* ☞ For $y = c\sqrt[n]{ax+b} + k$ with n even, domain: $ax + b \geq 0$.

2. *Range of Even Root Radical Functions* ☞ For $y = c\sqrt[n]{ax+b} + k$ with n even, if $c > 0$, then range: $y \geq k$. If $c < 0$, then range: $y \leq k$.

3. *Domain and Range of Odd Root Radical Functions* ☞ For $y = c\sqrt[n]{ax+b} + k$ with n odd, domain: All real numbers ($\mathbb{R}$). Range: All real numbers ($\mathbb{R}$).

Simplifying Radicals with Fractional Components

1. *Simplification of Fractional Radicals* ☞ $\sqrt[n]{\frac{a}{b}} = \frac{\sqrt[n]{a}}{\sqrt[n]{b}}$ where $b \neq 0$.

 **Rational Expressions**

Graphing Rational Functions

1. *Rational Function Form* ☞ $f(x) = \frac{p(x)}{q(x)}$ where $p(x)$ and $q(x)$ are polynomials and $q(x) \neq 0$.
2. *Domain of a Rational Function* ☞ The domain is all real numbers except the roots of $q(x)$.
3. *Vertical Asymptote* ☞ Vertical line $x = a$ where a is a root of $q(x)$.
4. *Horizontal Asymptote* ☞ Horizontal line $y = b$ where $b = \frac{p(n)}{q(n)}$ for large n ($n \to \pm\infty$), if degree of $p(x) \leq$ degree of $q(x)$.

Number Sequences and Formulas

Evaluating Variable Expressions for Number Sequences

1. *General Form of a Sequence* ☞ $A_n = f(n)$ where A_n is the n-th term and $f(n)$ is a function representing the formula of the sequence.

Writing Variable Expressions for Arithmetic Sequences

1. *General Form of an Arithmetic Sequence* ☞ $A_n = a + (n-1)d$ where a is the first term, d is the common difference, and n is the term number.

Writing Variable Expressions for Geometric Sequences

1. *Expression for the n-th Term of a Geometric Sequence* ☞ $G_n = a \times r^{n-1}$ where a is the first term, r is the common ratio, and n is the term number.

Formulating a Recursive Sequence

1. *Recursive Formula for Arithmetic Sequence* ☞ $a_n = a_{n-1} + d$ where d is the common difference.
2. *Recursive Formula for Geometric Sequence* ☞ $a_n = a_{n-1} \times r$ where r is the common ratio.
3. *Recursive Formula for Fibonacci Sequence* ☞ $a_n = a_{n-1} + a_{n-2}$ for $n \geq 3$, with $a_1 = a_2 = 1$.

Formulas for Chapter: Direct and Inverse Variations

Determining the Constant of Variation

1. *Direct Variation Formula* 👉 $y = kx$ where $k = \frac{y}{x}$ and k is the constant of variation.
2. *Inverse Variation Formula* 👉 $xy = k$ where k can be found using $k = xy$ and k is the constant of variation.

Formulas for Chapter: Statistical Analysis and Probability

Calculating Mean, Median, Mode, and Range

1. *Mean (Average)* 👉 Mean $= \frac{x_1 + x_2 + \cdots + x_n}{n}$
2. *Median* 👉 Order data. For odd n, median = middle value. For even n, median = average of two middle values.
3. *Mode* 👉 Most frequently occurring value(s) in the data set.
4. *Range* 👉 Range = Maximum − Minimum

Creating a Pie Graph

1. *Calculating the Angle of a Pie Slice* 👉 angle $= \frac{\text{data value}}{\text{total data sum}} \times 360°$

Calculating and Interpreting Correlation Coefficients

1. *Sample Standard Deviation* 👉 $s_x = \sqrt{\frac{1}{n-1} \sum_{i=1}^{n} (x_i - \bar{x})^2}$
2. *Sample Pearson Correlation Coefficient* 👉 $r = \frac{1}{n-1} \sum_{i=1}^{n} \left(\frac{x_i - \bar{x}}{s_x}\right)\left(\frac{y_i - \bar{y}}{s_y}\right)$

Formulating the Equation of a Regression Line

1. *Regression Line Equation* 👉 $y = mx + b$ where $m = r \times \frac{s_y}{s_x}$ and $b = \bar{y} - m\bar{x}$.

Understanding Permutations and Combinations

1. *Permutations Formula* 👉 $P(n,r) = \frac{n!}{(n-r)!}$, where $n!$ denotes the factorial of n.
2. *Combinations Formula* 👉 $C(n,r) = \frac{n!}{r!(n-r)!}$

Solving Probability Problems

1. *Basic Probability Formula* 👉 $P(A) = \frac{\text{number of desired outcomes}}{\text{total number of possible outcomes}}$
2. *Probability of Independent Events* 👉 $P(A \text{ and } B) = P(A) \times P(B)$ where A and B are independent events.

15. Practice Test 1

15.1 Practices

1) Which statement is incorrect for the function $f(x) = 4x^2 - 8x + 3$?

   ☐ A. The axis of symmetry of the function f is $x = 1$.

   ☐ B. The vertex of the function f is at $(1, -1)$.

   ☐ C. The zeros of f are 1 and $\frac{3}{4}$.

   ☐ D. The function opens upwards.

2) Which of the following is a factor of $8x^{10} - 14x^5 + 2x^4$?

   ☐ A. $2x^6 - 7x + 1$

   ☐ B. $x - 2$

   ☐ C. $2x^2 + 1$

   ☐ D. $4x^6 - 7x + 1$

3) Determine the equation of a horizontal line passing through the point $(4, -7)$.

   ☐ A. $y = -7$

   ☐ B. $x = -7$

   ☐ C. $y = 4$

   ☐ D. $x = 4$

4) A segment of an exponential function is depicted on a coordinate plane.
   Which of the following correctly describes the domain of the segment shown?

☐ A. $y \geq 7$

☐ B. $-\infty < x < 3$

☐ C. $x \leq 2$

☐ D. $0 < y < 7.2$

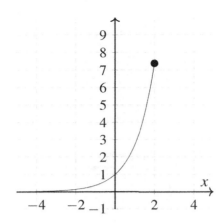

5) What is the product of $(6x+3y)(4x+3y)$?

☐ A. $24x^2 + 30xy + 9y^2$

☐ B. $3x^2 + 6xy + 3y^2$

☐ C. $6x^2 + 18xy + 3y^2$

☐ D. $10x^2 + 15xy + 9y$

6) Which of the following functions is equivalent to $g(x) = 6x^2 - 36x - 5$?

☐ A. $g(x) = 6(x-3)^2 + 50$

☐ B. $g(x) = 6(x+3)^2 - 50$

☐ C. $g(x) = 6(x-3)^2 - 59$

☐ D. $g(x) = 6(x-50)^2 - 3$

7) Among the given representations, which one shows y as a function of x?

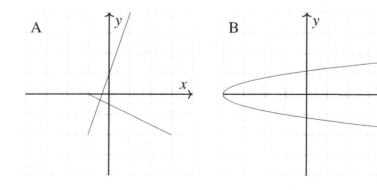

15.1 Practices

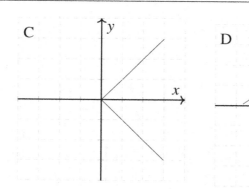

- ☐ A. Graph A
- ☐ B. Graph B
- ☐ C. Graph C
- ☐ D. Graph D

8) Which graph best represents the solution set of $y > \frac{1}{2}x - 2$?

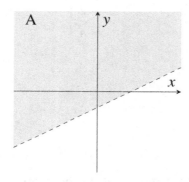 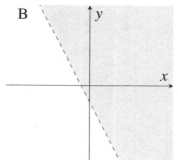

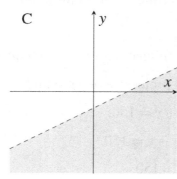 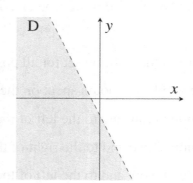

- ☐ A. Graph A
- ☐ B. Graph B
- ☐ C. Graph C
- ☐ D. Graph D

9) Given two characteristics of a quadratic function g:

- The axis of symmetry of the graph of g is $x = -2$.
- Function g has exactly one zero.

Which graph could represent g?

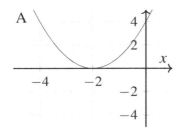
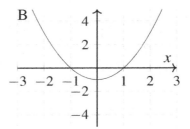
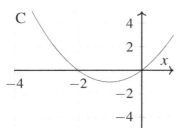
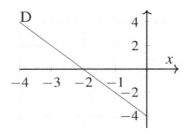

☐ A. Graph A

☐ B. Graph B

☐ C. Graph C

☐ D. Graph D

10) A freelancer works on two different projects. The total time spent on these projects cannot exceed 150 hours in a month.

Which option describes the solution set for all possible combinations of x, the number of hours spent on the first project, and y, the number of hours spent on the second project, in one month?

☐ A. A shaded region below and to the left of the line $x + y = 150$.

☐ B. A shaded region above and to the right of the line $x + y = 150$.

☐ C. A shaded region below and to the left of the line $x = 150$.

☐ D. A shaded region below and to the left of the line $y = 150$.

11) A blog's monthly views and comments are recorded in a table, indicating a linear relationship. Based on the data, what is the best prediction of the number of comments for each post if the number of monthly views reaches 6,000?

15.1 Practices

☐ A. 75
☐ B. 90
☐ C. 110
☐ D. 125

Monthly Views	Comments
3600	70
2400	50
4800	90
1200	30

12) A graph depicts the linear relationship between the volume of water in a tank and the time it has been filling.

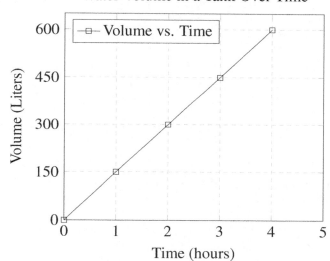

Which of these best represents the rate of change of the water volume with respect to the time elapsed?

☐ A. $150\frac{L}{hr}$

☐ B. $\frac{1}{150}\frac{L}{hr}$

☐ C. $250\frac{L}{hr}$

☐ D. $\frac{1}{250}\frac{L}{hr}$

13) The graph of $h(x) = x^3$ is transformed to create the graph of $k(x) = 0.5h(x)$. Which graph best represents h and k?

Option A

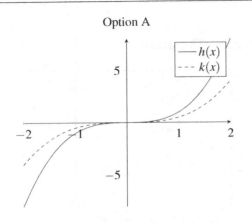

Option B

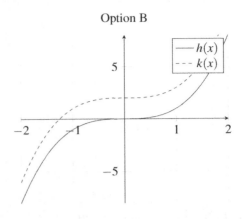

Option C

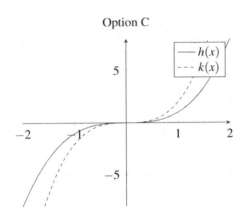

Option D

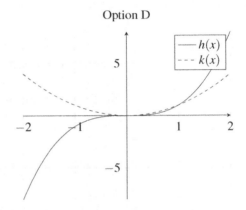

14) What is the value of the y-intercept of the graph $g(x) = 8.5(0.7)^x$? Write your answer in the box:

15) What is the ratio of the maximum value to the minimum value of the function $h(x) = -2x + 4$, where $-1 \leq x \leq 4$?

- A. $-\frac{5}{2}$
- B. $-\frac{3}{2}$
- C. $-\frac{10}{7}$
- D. $\frac{8}{3}$

16) A system of linear equations is represented by lines u and v. A table representing some points on line u and the graph of line v are provided.

15.1 Practices

Line u	
x	y
0	2
2	3
4	4

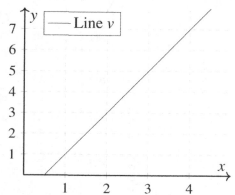

Which system of equations is best represented by lines u and v?

- ☐ A. $3x - 2y = 1$
     $x + y = 2$

- ☐ B. $3x - 2y = 1$
     $x - 2y = 4$

- ☐ C. $y = \frac{2}{3}x - 2$
     $y = 2 - \frac{1}{3}x$

- ☐ D. $y = \frac{1}{2}x + 2$
     $y = 2x - 1$

17) In 2005, a programmer's income increased by $1,500 per year starting from a $35,000 annual salary. Which equation represents income greater than average?

(Let J represent income, y represent the number of years after 2005)

- ☐ A. $J > 1,500y + 35,000$
- ☐ B. $J > -1,500y + 35,000$
- ☐ C. $J < -1,500y + 35,000$
- ☐ D. $J < 1,500y - 35,000$

18) Solve: $\frac{4x+8}{x+4} \times \frac{x+4}{x+2} =$.

- ☐ A. 1
- ☐ B. 2
- ☐ C. 3
- ☐ D. 4

19) A graph shows the growth of a plant in height over several days.

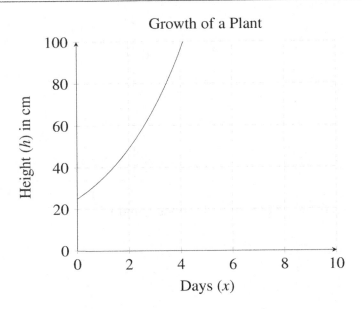

Based on this information, which function best describes the height of the plant h in centimeters per day?

☐ A. $h = 40(0.75)^x$

☐ B. $h = 25(1.4)^x$

☐ C. $h = 1.4(25)^x$

☐ D. $h = 0.75(40)^x$

20) Solve the following equation: $3x + 7 = 4x - 8$. Write the answer in the below box.

21) Simplify the expression $\frac{8}{\sqrt{18}-4}$.

☐ A. $\sqrt{18} + 4$

☐ B. 3

☐ C. $4\sqrt{18} + 16$

☐ D. $4\sqrt{18}$

22) A golfer hits a ball, and the graph shows the height in meters of the golf ball above the ground as a quadratic function of d, the horizontal distance in meters from the golfer.

15.1 Practices

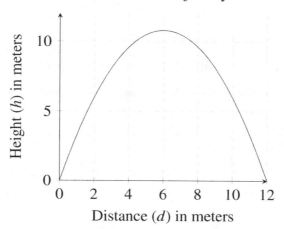

Golf Ball Trajectory

What is the domain of the function for this scenario?

☐ A. $0 \leq d \leq 12$

☐ B. $0 \leq h \leq 12$

☐ C. $2 \leq h \leq 3.5$

☐ D. $2 \leq d \leq 12$

23) Which of the following numbers is NOT a solution to the inequality $4x - 7 \leq 5x + 2$?

☐ A. -9

☐ B. -5

☐ C. -3

☐ D. -10

24) Which expression is equivalent to $16m^2 - 64$?

☐ A. $(4m - 8)(4m - 8)$

☐ B. $16(m - 4)$

☐ C. $16m(m - 4)$

☐ D. $16(m - 2)(m + 2)$

25) What is the value of the y-intercept of the graph of $h(x) = 15(1.3)^{x+2}$?

☐ A. 11.7

☐ B. 15

☐ C. 19.5

☐ D. 25.35

26) A nutritionist monitored the diet plan for a child for 24 weeks. The table and scatterplot show the calorie intake as a percentage of the child's daily requirement as a linear function of their age in weeks. What is the best prediction of the percentage of the daily calorie requirement that should be considered in the diet plan when the child is 30 weeks old?

Age (Weeks)	Calorie Intake (%)
0	100
4	102
8	105
12	107
16	109
20	112
24	115

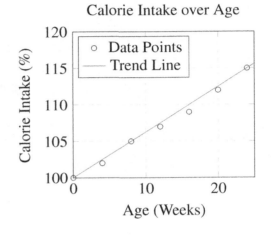

☐ A. 118%

☐ B. 120%

☐ C. 122%

☐ D. 125%

27) The graph of quadratic function h is displayed on a coordinate plane. What is the y-intercept of the graph of h? Write the answer in the below box.

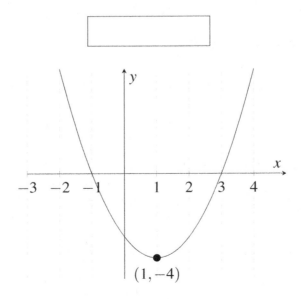

15.1 Practices

28) The line graphed on a coordinate plane represents the first of two equations in a system of linear equations. If the graph of the second equation in the system passes through the points $(0,-3)$ and $(4,5)$, which statement is true?

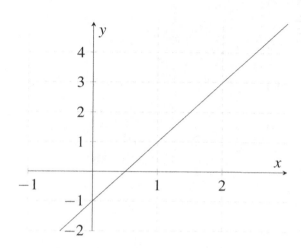

- ☐ A. The only solution to the system is $(-4,0)$.
- ☐ B. The only solution to the system is $(3,2)$.
- ☐ C. The system has no solution.
- ☐ D. The system has an infinite number of solutions.

29) Given $f(x) = x^2 - 7x + 12$, which statement is true about the zeroes of f?

- ☐ A. The zeroes are 4 and -3 because the factors of f are $(x-4)$ and $(x+3)$.
- ☐ B. The zeroes are -4 and 3 because the factors of f are $(x+4)$ and $(x-3)$.
- ☐ C. The zeroes are 4 and 3 because the factors of f are $(x-4)$ and $(x-3)$.
- ☐ D. The zeroes are -4 and -3 because the factors of f are $(x+4)$ and $(x+3)$.

30) A study shows the relationship between the amount of fertilizer used (in kilograms) and the yield of a crop (in tons). The data, represented by a quadratic function, is gathered over several seasons. Which function could best model this data?

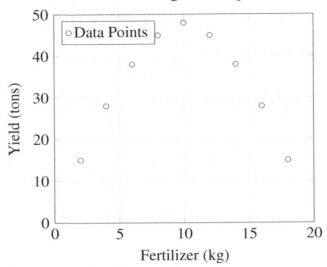

- ☐ A. $y = -x^2 + 20x$
- ☐ B. $y = x^2 - 20x$
- ☐ C. $y = -0.5x^2 + 10x$
- ☐ D. $y = 0.5x^2 - 10x$

31) Four tables display the relationship between temperature (in Celsius) and the solubility of a substance (in grams per 100ml of water). Which table does NOT represent solubility as a function of temperature?

☐ A.

Temperature (°C)	Solubility (g/100ml)
5	12
10	15
15	18
20	21

☐ B.

Temperature (°C)	Solubility (g/100ml)
5	14
10	16
15	18
20	20

15.1 Practices

	Temperature (°C)	Solubility (g/100ml)
C.	5	15
	5	16
	10	20
	15	24

	Temperature (°C)	Solubility (g/100ml)
D.	5	13
	10	17
	15	21
	20	25

32) If 40% of z is equal to 20% of 30, what is the value of $(z+5)^2$?

☐ A. 225

☐ B. 400

☐ C. 350

☐ D. 600

33) Consider a function $h(x)$ that has three distinct zeros. Which of the following could represent the graph of $h(x)$?

☐ A. A cubic graph crossing the x-axis at three different points.

☐ B. A quadratic graph touching the x-axis at one point and crossing at another.

☐ C. A linear graph crossing the x-axis at one point.

☐ D. A quartic graph touching the x-axis at two points and crossing at two others.

34) Determine the negative solution to the equation $2x^3 + 5x^2 - 12x = 0$.

35) A portion of a linear function h is graphed on a coordinate plane.

Which inequality best represents the domain of the part shown?

- A. $-3 < x \leq 3$
- B. $-3 \leq x < 3$
- C. $-1 < h(x) \leq 5$
- D. $-1 \leq h(x) < 5$

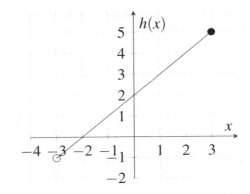

36) A linear function f has an x-intercept of 2 and a y-intercept of -3. Which option best describes f?
    - A. A line crossing the x-axis at 2 and y-axis at -3.
    - B. A line crossing the x-axis at -3 and y-axis at 2.
    - C. A line with a positive slope crossing the x-axis at 2 and y-axis at -3.
    - D. A line with a negative slope crossing the x-axis at 2 and y-axis at -3.

37) Given the system of equations: $4x + 3y = 7$ and $2x - 6y = -4$, which of the following ordered pairs (x,y) satisfies both equations?
    - A. $(2,1)$
    - B. $(0,-1)$
    - C. $(3,-2)$
    - D. $(1,1)$

38) A line in the xy-plane passes through the point $(1,2)$ and has a slope of $\frac{1}{2}$. Which of the following points lies on this line?
    - A. $(3,4)$
    - B. $(4,1.5)$
    - C. $(7,4.5)$
    - D. $(5,4)$

39) What is the simplified form of $(5m^2 + 3m + 4) - (3m^2 - 6)$?
    - A. $2m^2 + 3m + 10$
    - B. $2m^2 + 3m - 2$
    - C. $2m + 10$
    - D. $m^2 + 3m - 2$

15.1 Practices

40) Solve for x: $4(x+1) = 6(x-4)+20$.

☐ A. 0

☐ B. 2

☐ C. 4

☐ D. 6.5

41) The given expression $2x-3$ is a factor of which equation?

☐ A. $(2x-3)+17$

☐ B. $2x^3 - x^2 - 3x$

☐ C. $6(3+2x)$

☐ D. $2x^2 + x - 6$

42) What is the number of solutions to the equation $x^2 - 3x + 1 = x - 3$?

☐ A. 0

☐ B. 1

☐ C. 2

☐ D. Infinite

43) In the xy-plane, if $(0,0)$ is a solution to the system of inequalities $y < c - x$ and $y > x + b$, which of the following relationships between c and b must be true?

☐ A. $c < b$

☐ B. $c > b$

☐ C. $c = b$

☐ D. $c = b + c$

44) Calculate $f(5)$ for the following function f:

$$f(x) = x^2 - 3x.$$

☐ A. 5

☐ B. 10

☐ C. 15

☐ D. 20

45) John buys a pepper plant that is 5 inches tall. With regular watering, the plant grows 3 inches a year. Writing John's plant's height as a function of time, what does the y-intercept represent?

☐ A. The y-intercept represents the rate of growth of the plant which is 5 inches.

☐ B. The y-intercept represents the starting height of 5 inches.

☐ C. The y-intercept represents the rate of growth of the plant which is 3 inches per year.

☐ D. The y-intercept is always zero.

46) Multiply and write the product in scientific notation:

$$(3.1 \times 10^7) \times (1.8 \times 10^{-4}).$$

☐ A. 5.58×10^3

☐ B. 5.58×10^{11}

☐ C. 55.8×10^2

☐ D. 5.58×10^2

47) The first five terms in a geometric sequence are shown, where $x_1 = 3$.

3, −6, 12, −24, 48.

Based on this information, which equation can be used to find the nth term in the sequence, x_n?

☐ A. $x_n = 3(-2)^{n-1}$

☐ B. $x_n = 3^n$

☐ C. $x_n = -3n$

☐ D. $x_n = 3(-1)^n$

48) What is the parent graph of the following function and what transformations have taken place on it? $y = 2x^2 - 8x + 6$

☐ A. The parent graph is $y = x^2$, which is stretched vertically by a factor of 2 and shifted 2 units right and 2 units down.

☐ B. The parent graph is $y = x^2$, which is shifted 4 units right and 6 units up.

☐ C. The parent graph is $y = x^2$, which is stretched vertically by a factor of 2 and shifted 4 units left.

☐ D. The parent graph is $y = x^2$, which is compressed vertically and shifted 2 units right and 6 units up.

49) A biology research group observes the growth of a certain algae species in a lake. The table shows the

15.1 Practices

measured algae coverage area (in square meters) over several years since 2010. The data can be modeled by an exponential function.

Which function best models the data?

- A. $a(x) = 500(1.08)^x$
- B. $a(x) = 750(0.95)^x$
- C. $a(x) = 1.08(500)^x$
- D. $a(x) = 0.95(750)^x$

Year (since 2010)	Algae Coverage (sq. meters)
0	500
1	540
2	583.2
3	629.8
4	680.6

50) Determine the value of x in the solution of the system of equations:

$$4x + 5y = 17$$
$$2x - 3y = 3$$

- A. 1
- B. 2
- C. 3
- D. 4

51) Solve the following inequality: $|2x+4| \leq 6$.
- A. $x \leq 1$ or $x \geq -5$
- B. $-5 \leq x \leq 1$
- C. $x \leq -5$ or $x \geq 1$
- D. $-1 \leq x \leq 5$

52) How many ways can we pick a team of 4 people from a group of 10?
- A. 120
- B. 210
- C. 252
- D. 5040

53) The average of 7, 12, 25, and y is 15. What is the value of y?

☐ A. 11

☐ B. 20

☐ C. 15

☐ D. 16

54) In triangle DEF, if the measure of angle D is 45 degrees, what is the value of y? (Note: The figure is not drawn to scale)

☐ A. 40

☐ B. 50

☐ C. 55

☐ D. 58

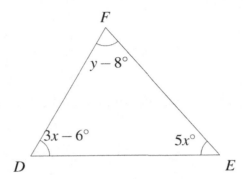

55) An angle is equal to one fifth of its complement. What is the measure of that angle?

☐ A. 18

☐ B. 10

☐ C. 15

☐ D. 36

56) When a number is subtracted from 35 and the difference is divided by that number, the result is 4. What is the value of the number?

☐ A. 3

☐ B. 5

☐ C. 7

☐ D. 9

57) Which of the following expressions is the inverse of the function $f(x) = \sqrt{x} + 2$?

☐ A. $(x+2)^2$

☐ B. $x^2 - 4x$

☐ C. $x^2 - 4x + 4$

☐ D. $x^2 - 4x + 2$

15.1 Practices

58) A chemical solution contains 8% alcohol. If there is 32 ml of alcohol, what is the volume of the solution?

☐ A. 300 ml

☐ B. 400 ml

☐ C. 500 ml

☐ D. 800 ml

59) What is the value of y in this equation?

$$5\sqrt{2y} + 7 = 27$$

☐ A. 4

☐ B. 20

☐ C. 50

☐ D. 8

60) A function $h(4) = 3$ and $h(6) = 5$. A function $k(5) = 7$ and $k(3) = 8$. What is the value of $k(h(6))$?

☐ A. 7

☐ B. 8

☐ C. 5

☐ D. 3

15.2 Answer Keys

1) C. The zeros of f are 1 and $\frac{3}{4}$.
2) D. $4x^6 - 7x + 1$
3) A. $y = -7$
4) C. $x \leq 2$
5) A. $24x^2 + 30xy + 9y^2$
6) C. $g(x) = 6(x-3)^2 - 59$
7) D. Graph D
8) A. Graph A
9) A.
10) A. A shaded region below and to the left of the line $x + y = 150$.
11) C. 110
12) A. $150\frac{L}{hr}$
13) A.
14) 8.5
15) B. $-\frac{3}{2}$
16) D. $\begin{array}{l} y = \frac{1}{2}x + 2 \\ y = 2x - 1 \end{array}$
17) A. $J > 1,500y + 35,000$
18) D. 4
19) B. $h = 25(1.4)^x$
20) 15
21) C. $4\sqrt{18} + 16$
22) A. $0 \leq d \leq 12$
23) D. -10
24) D. $16(m-2)(m+2)$
25) D. 25.35
26) A. 118%
27) -3
28) C. The system has no solution.
29) C.
30) C. $y = -0.5x^2 + 10x$
31) C.
32) B. 400
33) A.
34) -4
35) A. $-3 < x \leq 3$
36) C. A line with a positive slope crossing the x-axis at 2 and y-axis at -3.
37) D. $(1,1)$
38) D. $(5,4)$
39) A. $2m^2 + 3m + 10$
40) C. 4
41) D. $2x^2 + x - 6$
42) B. 1
43) B. $c > b$
44) B. 10
45) B.
46) A. 5.58×10^3
47) A. $x_n = 3(-2)^{n-1}$
48) A.
49) A. $a(x) = 500(1.08)^x$
50) C. 3
51) B. $-5 \leq x \leq 1$
52) B. 210
53) D. 16

15.2 Answer Keys

54) D. 58

55) C. 15

56) C. 7

57) C. $x^2 - 4x + 4$

58) B. 400 ml

59) D. 8

60) A. 7

15.3 Answers with Explanation

1) The function $f(x) = 4x^2 - 8x + 3$ can be rewritten as $f(x) = 4(x-1)^2 - 1$.

$$\text{Axis of symmetry} = 1 \Rightarrow \text{A is correct.}$$

$$\text{Vertex} = (1, -1) \Rightarrow \text{B is correct.}$$

$$a = 4 > 0 \quad \Rightarrow \text{Opens upwards} \Rightarrow \text{D is correct.}$$

The zeros of the function are found by solving $4x^2 - 8x + 3 = 0$. By the quadratic formula, the zeros (which are $\frac{1}{2}$ and $\frac{3}{2}$), do not match the values in option C.

Thus, option C is incorrect.

2) The polynomial $8x^{10} - 14x^5 + 2x^4$ can be factored as:

$$2x^4(4x^6 - 7x + 1).$$

Therefore, D is the correct choice.

3) A horizontal line has the form $y = k$, where k is a constant. The given point is $(4, -7)$, so the line must have a constant y-value of -7. Thus, the equation of the line is $y = -7$, which corresponds to option A.

4) The domain of a function refers to the set of all possible x values for which the function is defined. In the graph, the exponential function segment is shown continuing leftwards to $-\infty$ and ending at $x = 2$, where it is marked by a filled circle. This indicates that the function is defined for all x values up to and including 2. Therefore, the correct description of the domain of the segment shown is $x \leq 2$, as it includes all x values less than or equal to 2, making option C correct.

5) Multiplying the polynomials $(6x + 3y)$ and $(4x + 3y)$ gives:

$$(6x + 3y)(4x + 3y) = 24x^2 + 18xy + 12xy + 9y^2 = 24x^2 + 30xy + 9y^2.$$

Therefore, option A is the correct answer.

15.3 Answers with Explanation

6) To find the equivalent form of $g(x) = 6x^2 - 36x - 5$, complete the square:

$$g(x) = 6(x^2 - 6x) - 5 = 6\left[(x-3)^2 - 3^2\right] - 5 = 6(x-3)^2 - 6 \times 9 - 5.$$

Simplifying gives:

$$g(x) = 6(x-3)^2 - 54 - 5 = 6(x-3)^2 - 59.$$

Therefore, option C is the correct equivalent function.

7) To determine which graph represents y as a function of x, we need to check if each x value corresponds to exactly one y value. This is commonly known as the "vertical line test," where a vertical line drawn at any x-coordinate should intersect the graph at no more than one point.

- Graph A shows two linear functions intersecting, meaning at some x values, there are two corresponding y values. It fails the vertical line test.

- Graph B depicts a sideways parabola. This graph fails the vertical line test because a vertical line can intersect the graph at two points in certain areas.

- Graph C shows two intersecting linear functions, similar to Graph A, and also fails the vertical line test.

- Graph D represents a square root function. Every x value in its domain corresponds to exactly one y value, successfully passing the vertical line test.

Therefore, Graph D is the only one that correctly represents y as a function of x, making option D correct.

8) To determine which graph represents the solution set of the inequality $y > \frac{1}{2}x - 2$, we need to identify the graph that correctly depicts this relationship.

- Graph A shows the line $y = \frac{1}{2}x - 2$ as a dashed line, indicating that the line itself is not included in the solution set. The shaded area above the line represents all the (x, y) points where y is greater than $\frac{1}{2}x - 2$. This correctly represents the solution set of the given inequality.

- Graph B depicts a line with a negative slope and a different inequality, which is not relevant to the given inequality $y > \frac{1}{2}x - 2$.

- Graph C, similar to Graph A, shows the line $y = \frac{1}{2}x - 2$, but the shading is below the line, representing $y < \frac{1}{2}x - 2$, which is not the inequality we are looking to represent.

- Graph D, like Graph B, represents a different inequality not related to $y > \frac{1}{2}x - 2$.

Therefore, Graph A correctly represents the solution set for $y > \frac{1}{2}x - 2$, making option A correct.

9) To determine which graph could represent the quadratic function g with the given characteristics, we need to analyze the provided graphs based on two criteria: the axis of symmetry and the number of zeros.

- The axis of symmetry of a quadratic function is a vertical line that divides the graph into two mirror-image halves. For function g, the axis of symmetry is given as $x = -2$.

Other options have not both of these properties.

Therefore, Graph A correctly matches both the axis of symmetry at $x = -2$ and the presence of exactly one zero, making option A the correct choice.

10) The inequality representing the condition is $x + y \leq 150$. This inequality means the combined hours x and y should not exceed 150. Therefore, the solution set is represented by the region below and to the left of the line $x + y = 150$, which is option A.

11) Despite the rearrangement, the relationship between monthly views and comments is still linear. To find the number of comments for 6,000 views, we first calculate the slope (m) of the linear relationship.

Using two points from the table, for example, (3600, 70) and (1200, 30), we find the slope:

$$m = \frac{70 - 30}{3600 - 1200} = \frac{40}{2400} = \frac{1}{60}.$$

The linear equation is $y = mx + b$. Using the point (1200, 30) to solve for 'b':

$$30 = \frac{1}{60} \times 1200 + b \Rightarrow b = 30 - 20 = 10.$$

Thus, the equation is $y = \frac{1}{60}x + 10$. For 6,000 views:

$$y = \frac{1}{60} \times 6000 + 10 = 100 + 10 = 110.$$

Therefore, the correct answer is option C.

12) The rate of change of the water volume with respect to time can be determined by calculating the slope of the line on the graph. The slope represents the change in volume per unit time, which is the rate at which the tank is being filled.

From the graph, we can take two points to calculate the slope. For instance, using the points (1 hr, 150 L)

15.3 Answers with Explanation

and (2 hrs, 300 L):

$$\text{Slope} = \frac{\text{Change in Volume}}{\text{Change in Time}} = \frac{300L - 150L}{2\,hr - 1\,hr} = \frac{150L}{1\,hr} = 150\frac{L}{hr}.$$

Therefore, the rate of change of the water volume with respect to the time elapsed is $150\frac{L}{hr}$, making option A correct.

13) The transformation from $h(x) = x^3$ to $k(x) = 0.5h(x)$ involves scaling the graph of $h(x)$ vertically by a factor of 0.5. This means that the heights of the points on $h(x)$ are halved to get the corresponding points on $k(x)$, resulting in a vertically scaled (compressed) version of the cubic function. Therefore, option A is correct.

14) The y-intercept of a function is the value of y when $x = 0$. For the function $g(x) = 8.5(0.7)^x$, substituting $x = 0$ gives:

$$g(0) = 8.5(0.7)^0 = 8.5 \times 1 = 8.5.$$

Therefore, the y-intercept of the graph is 8.5.

15) For the function $h(x) = -2x + 4$, evaluate at the endpoints of the interval:

$$h(-1) = -2(-1) + 4 = 2 + 4 = 6 \quad \text{(Maximum value)},$$

$$h(4) = -2(4) + 4 = -8 + 4 = -4 \quad \text{(Minimum value)}.$$

The ratio of the maximum value to the minimum value is:

$$\frac{\text{Maximum}}{\text{Minimum}} = \frac{6}{-4} = -\frac{3}{2}.$$

Therefore, the correct answer in option B.

16) To determine which system of equations is best represented by lines u and v, we need to derive the equations for these lines based on the given information.

- For line u, we can use the points from the table to find its equation. With points $(0, 2)$ and $(2, 3)$, the slope m is:

$$m = \frac{3 - 2}{2 - 0} = \frac{1}{2}.$$

Using the point-slope form $y - y_1 = m(x - x_1)$ with point $(0, 2)$:

$$y - 2 = \frac{1}{2}(x - 0) \Rightarrow y = \frac{1}{2}x + 2.$$

- For line v, the equation is provided in the graph's legend: $y = 2x - 1$.

Comparing these derived equations with the provided options, we find that option D, which states:

$$y = \tfrac{1}{2}x + 2$$
$$y = 2x - 1$$

perfectly matches the equations for lines u and v, making it the correct choice.

17) The equation for the programmer's income as a function of years after 2005 is $J = 1,500y + 35,000$, where $1,500$ is the annual increase and $35,000$ is the starting salary. To represent income greater than the average, the inequality would be $J > 1,500y + 35,000$, making option A correct.

18) Simplify the expression:
$$\frac{4x+8}{x+4} \times \frac{x+4}{x+2} = \frac{4(x+2)}{x+4} \times \frac{x+4}{x+2}.$$

The factors $x+4$ and $x+2$ cancel out, leaving: 4, option D.

19) The graph of the function $h = 25(1.4)^x$ shows a pattern of exponential growth, which is typical for the height of a plant over time. In this function, 25 represents the initial height of the plant, and 1.4 is the growth factor per day. As x (days) increases, the height h increases at an exponential rate, indicating that the plant grows faster as time progresses. This type of growth is common in many biological processes, making Option B the best representation of the plant's growth. The other options either depict a decreasing growth pattern or an unrealistic exponential increase.

20) To solve the equation $3x + 7 = 4x - 8$, we first move all x terms to one side and the constants to the other side:

$$3x - 4x = -8 - 7$$

$$-x = -15 \Rightarrow x = \frac{-15}{-1} = 15.$$

Therefore, the solution of the equation is $x = 15$.

15.3 Answers with Explanation

21) To simplify $\frac{8}{\sqrt{18}-4}$, use the conjugate to rationalize the denominator:

$$\frac{8}{\sqrt{18}-4} \times \frac{\sqrt{18}+4}{\sqrt{18}+4} = \frac{8(\sqrt{18}+4)}{18-16}.$$

Simplifying further:

$$\frac{8\sqrt{18}+32}{2} = 4\sqrt{18}+16.$$

Therefore, the correct answer is option C.

22) The graph of the quadratic function representing the golf ball's trajectory shows the height of the ball as it travels horizontally. The domain of this function, d, represents the horizontal distance the ball covers from the golfer. In this scenario, the domain is given as $0 \leq d \leq 12$ meters, which is Option A. The other options (B, C, D) refer to the range of heights or incorrect distance ranges, which are not appropriate to describe the domain of the function in this context.

23) Solving the inequality $4x - 7 \leq 5x + 2$:

$$4x - 7 \leq 5x + 2 \Rightarrow -7 - 2 \leq 5x - 4x \Rightarrow -9 \leq x.$$

Therefore, x must be greater than or equal to -9. The number -10 does not satisfy this condition, making option D the correct choice.

24) The expression $16m^2 - 64$ is a difference of squares, which can be factored as:

$$16m^2 - 64 = 16(m^2 - 4) = 16(m-2)(m+2).$$

Therefore, option D is the correct equivalent expression.

25) The y-intercept of a function is found by setting $x = 0$. For the function $h(x) = 15(1.3)^{x+2}$:

$$h(0) = 15(1.3)^{0+2} = 15 \times 1.3^2 = 15 \times 1.69 = 25.35.$$

Therefore, the correct answer is option D.

26) The scatter plot shows a linear trend in the child's calorie intake as a percentage of their daily requirement over time. The trend line approximates this relationship, and we can use it to predict the calorie intake at 30 weeks. Based on the trend line's equation derived from the scatter plot, the best prediction for 30 weeks can be calculated. If the trend line equation is approximately $y = 100 + 0.625x$, then at $x = 30$ weeks, the prediction is $y = 100 + 0.625 \times 30 = 118.75\%$. This value is closest to Option A, 118%.

27) To determine the y-intercept of the graph of the quadratic function h, which we do not have in explicit form, we can utilize the given points on the graph: $(1, -4)$, $(-1, 0)$, and $(3, 0)$. The y-intercept occurs where the graph crosses the y-axis, which is at $x = 0$.

Since we know that the points $(-1, 0)$ and $(3, 0)$ are zeros of the function, the quadratic function h can be expressed in its factored form:

$$h(x) = a(x+1)(x-3),$$

where a is a constant.

To find a, we use the third point $(1, -4)$:

$$-4 = a(1+1)(1-3) = a \times 2 \times (-2) = -4a.$$

Solving for a gives us $a = 1$.

Now we have the function:

$$h(x) = (x+1)(x-3).$$

To find the y-intercept, substitute $x = 0$ into $h(x)$:

$$h(0) = (0+1)(0-3) = 1 \times (-3) = -3.$$

Therefore, the y-intercept of the graph is -3.

28) The first line's equation, as represented in the graph, appears to be $y = 2x - 1$.

The second line passes through the points $(0, -3)$ and $(4, 5)$. We can find its slope using these points:

$$\text{Slope} = \frac{5 - (-3)}{4 - 0} = \frac{8}{4} = 2.$$

15.3 Answers with Explanation

Since one of the points is $(0, -3)$, which is the y-intercept, the equation of the second line can be written as:

$$y = 2x - 3.$$

Comparing the two lines: - The first line is $y = 2x - 1$. - The second line is $y = 2x - 3$.

Both lines have the same slope but different y-intercepts. Therefore, they are parallel and do not intersect. When two lines in a system of linear equations are parallel and have different y-intercepts, they do not have any points of intersection, meaning there is no solution to the system. Hence, the correct answer is option C, indicating that the system has no solution.

29) To determine the zeroes of the quadratic function $f(x) = x^2 - 7x + 12$, we can factorize the quadratic equation. The zeroes are the values of x that make $f(x) = 0$.

The factorization of $f(x)$ involves finding two numbers that multiply to 12 (the constant term) and add up to -7 (the coefficient of the x term). These numbers are -4 and -3, as $-4 \times -3 = 12$ and $-4 + -3 = -7$. Therefore, the equation can be factored as:

$$f(x) = x^2 - 7x + 12 = (x - 4)(x - 3).$$

Setting each factor equal to zero gives the zeroes of the function:

$$x - 4 = 0 \Rightarrow x = 4,$$

$$x - 3 = 0 \Rightarrow x = 3.$$

Thus, the zeroes of the function are 4 and 3. These correspond to the factors $(x - 4)$ and $(x - 3)$. Therefore, the correct option is C: "The zeroes are 4 and 3 because the factors of f are $(x - 4)$ and $(x - 3)$."

30) The scatter plot shows a quadratic trend where the yield initially increases with the amount of fertilizer and then decreases after reaching a peak. Among the given options, the function $y = -0.5x^2 + 10x$ best models this trend. It has a positive linear term, indicating an initial increase in yield with increasing fertilizer, and a negative quadratic term, reflecting the decrease in yield beyond a certain point. Note that the option A is incorrect because for 10 kg of fertilizer, we have approximately 50 tons of the crop.

31) To determine which table does not represent solubility as a function of temperature, we need to identify a table where a single temperature value corresponds to more than one solubility value.

- Tables A, B, and D all have unique solubility values for each temperature, which is consistent with the definition of a function.

- Table C shows the temperature of 5°C corresponding to two different solubility values (15 and 16 g/100ml). This violates the definition of a function, where each input (temperature) should map to exactly one output (solubility).

Therefore, Table C is the one that does NOT represent solubility as a function of temperature.

32) Solve the equation $0.40z = 0.20 \times 30$:

$$0.40z = 6 \Rightarrow z = \frac{6}{0.40} = 15.$$

Then calculate $(z+5)^2$ with $z = 15$:

$$(15+5)^2 = 20^2 = 400.$$

Thus, the value of $(z+5)^2$ is 400.

33) A function with three distinct zeros will cross the x-axis at three separate points. This is typically characteristic of a cubic function, as cubic functions can have up to three real zeros. Therefore, a cubic graph that crosses the x-axis at three different points would represent $h(x)$.

34) To find the negative solution to the equation $2x^3 + 5x^2 - 12x = 0$, we first factor out the common term, which in this case is x:

$$x(2x^2 + 5x - 12) = 0.$$

The equation now has a factored term $x(2x^2 + 5x - 12)$. The roots of the equation are the solutions to $x = 0$ and $2x^2 + 5x - 12 = 0$. The solution $x = 0$ is already evident. To find the other solutions, we focus on the quadratic equation $2x^2 + 5x - 12 = 0$.

We can attempt to factorize the quadratic or use the quadratic formula. The quadratic formula is given by:

$$x = \frac{-b \pm \sqrt{b^2 - 4ac}}{2a},$$

15.3 Answers with Explanation

where $a = 2$, $b = 5$, and $c = -12$. Applying these values:

$$x = \frac{-5 \pm \sqrt{5^2 - 4 \times 2 \times -12}}{2 \times 2} = \frac{-5 \pm \sqrt{25 + 96}}{4} = \frac{-5 \pm \sqrt{121}}{4}.$$

This simplifies to:

$$x = \frac{-5 \pm 11}{4}.$$

The two solutions from this are:

$$x_1 = \frac{-5 + 11}{4} = \frac{6}{4} = 1.5,$$

$$x_2 = \frac{-5 - 11}{4} = \frac{-16}{4} = -4.$$

Among these, the negative solution is $x = -4$, which is the final answer.

35) The domain of a function represents the set of all possible input values (in this case, x values) for which the function is defined. In the graph, the portion of the linear function h shown runs from just beyond $x = -3$ to and including $x = 3$. This is represented by the inequality $-3 < x \leq 3$, where x is greater than -3 but less than or equal to 3. Therefore, the correct domain for the part of the function shown is Option A: $-3 < x \leq 3$.

36) The linear function f with an x-intercept of 2 and a y-intercept of -3 is best represented by a line that crosses the x-axis at 2 (point $(2,0)$) and the y-axis at -3 (point $(0,-3)$). Since the line moves from lower left to upper right, it has a positive slope. This matches Option C: A line with a positive slope crossing the x-axis at 2 and the y-axis at -3.

37) Solving the system by substitution or elimination method:

$$4x + 3y = 7$$
$$2x - 6y = -4$$

Multiply the second equation by -2 and add to the first equation:

$$4x + 3y = 7$$
$$-4x + 12y = 8$$

Adding these gives:
$$15y = 15 \Rightarrow y = 1.$$

Substitute y into one of the equations:
$$4x + 3(1) = 7 \Rightarrow x = 1.$$

Therefore, the ordered pair that satisfies both equations is $(1,1)$, which is option D.

38) Using the slope-point form of a line:
$$y - y_1 = m(x - x_1)$$

Substitute $(1,2)$ and slope $\frac{1}{2}$:
$$y - 2 = \frac{1}{2}(x - 1)$$

Simplify to get the equation of the line:
$$y = \frac{1}{2}x + \frac{3}{2}$$

Check each point to see which lies on the line. For $(5,4)$:
$$4 = \frac{1}{2}(5) + \frac{3}{2} = \frac{5}{2} + \frac{3}{2} = 4$$

Thus, point $(5,4)$ lies on the line, making option D correct.

39) Simplify the expression by subtracting the second polynomial from the first:
$$(5m^2 + 3m + 4) - (3m^2 - 6) = 5m^2 + 3m + 4 - 3m^2 + 6 = 2m^2 + 3m + 10.$$

Therefore, the simplified form is $2m^2 + 3m + 10$, which is option A.

40) Expand and simplify the equation:
$$4x + 4 = 6x - 24 + 20,$$

15.3 Answers with Explanation

$$4x + 4 = 6x - 4.$$

Bring all x terms to one side and constants to the other side:

$$4x - 6x = -4 - 4 \Rightarrow -2x = -8 \Rightarrow x = \frac{-8}{-2} = 4.$$

The solution is $x = 4$, making option C correct.

41) To determine if $2x - 3$ is a factor, it must divide one of the given equations without a remainder. We have:

$$2x^2 + x - 6 = (2x - 3)(x + 2).$$

Therefore, the correct answer in option D.

42) Rearrange the equation to form a quadratic equation:

$$x^2 - 3x + 1 = x - 3 \Rightarrow x^2 - 4x + 4 = 0.$$

Factorize or use the quadratic formula:

$$(x - 2)^2 = 0$$

The solution is $x = 2$ indicating only one unique solution, making option C correct.

43) Substituting $(0,0)$ into the inequalities:

$$0 < c - 0 \Rightarrow c > 0,$$

$$0 > 0 + b \Rightarrow b < 0.$$

Therefore, c must be greater than b for both inequalities to hold true at $(0,0)$, making option B correct.

44) Substitute $x = 5$ into the function:

$$f(5) = 5^2 - 3 \times 5 = 25 - 15 = 10.$$

Therefore, $f(5)$ equals 10, making option B correct.

45) In a linear function representing growth over time, the y-intercept represents the initial value of the variable being measured when time is zero. In this case, when time is zero (at the start), the height of the plant is 5 inches. Therefore, the y-intercept represents the starting height of the plant, which is 5 inches, making option B correct.

46) Multiply the numbers and add the exponents:

$$(3.1 \times 10^7) \times (1.8 \times 10^{-4}) = 5.58 \times 10^{7-4} = 5.58 \times 10^3.$$

Therefore, the product in scientific notation is 5.58×10^3, which is option A.

47) The sequence shows a pattern of each term being multiplied by -2 to get the next term. The first term is 3, and each subsequent term is obtained by multiplying the previous term by -2. The nth term of a geometric sequence is given by $x_1 \times r^{(n-1)}$, where x_1 is the first term and r is the common ratio. Thus, the formula for the nth term is:

$$x_n = 3(-2)^{n-1}.$$

Therefore, option A is correct.

48) The parent graph is $y = x^2$. The given function $y = 2x^2 - 8x + 6$ can be rewritten in vertex form as $y = 2(x-2)^2 - 2$. This indicates that the graph of $y = x^2$ has been transformed as follows:

1. Stretched vertically by a factor of 2 (due to the coefficient 2 in front of $(x-2)^2$).

2. Shifted 2 units to the right (due to the $(x-2)$ term).

3. Shifted 2 units down (due to the -2 outside the square).

Therefore, option A is correct.

49) The data in the table shows algae coverage increasing each year, which suggests exponential growth. In an exponential growth model, the value increases by a constant percentage each year. Option A ($a(x) = 500(1.08)^x$) represents an exponential function where the initial coverage is 500 sq. meters, and it grows by 8% each year (as indicated by the factor 1.08). This model aligns with the pattern observed in the table, where each year's coverage is approximately 8% more than the previous year.

15.3 Answers with Explanation

The other options either represent a decrease (due to factors less than 1) or an unrealistic exponential increase (with the base as 500 or 750). Therefore, Option A is the best model for the data.

50) To solve the system
$$4x + 5y = 17$$
$$2x - 3y = 3,$$
we can use the elimination method. Multiply the second equation by 2 to align the coefficients of x:
$$4x + 5y = 17$$
$$4x - 6y = 6.$$

Subtract the second equation from the first:
$$11y = 11.$$

This gives $y = 1$. Substitute $y = 1$ into the first equation:
$$4x + 5 = 17 \Rightarrow 4x = 12 \Rightarrow x = 3.$$

Thus, $x = 3$, option C.

51) To solve the inequality $|2x + 4| \leq 6$, we consider two cases based on the absolute value:

1. $2x + 4 \leq 6$
$$2x \leq 2 \Rightarrow x \leq 1.$$

2. $-(2x + 4) \leq 6$ which simplifies to $2x + 4 \geq -6$
$$2x \geq -10 \Rightarrow x \geq -5.$$

Combining both conditions, we get the solution set:
$$-5 \leq x \leq 1.$$

This is represented by option B.

52) To find the number of ways to pick a team of 4 people from a group of 10, we use the combination formula, which is given by:

$$\binom{n}{r} = \frac{n!}{r!(n-r)!},$$

where n is the total number of people, r is the number of people to choose, and ! denotes the factorial.

For our case, $n = 10$ and $r = 4$, so the formula becomes:

$$\binom{10}{4} = \frac{10!}{4!(10-4)!} = \frac{10 \times 9 \times 8 \times 7}{4 \times 3 \times 2 \times 1} = 210.$$

Therefore, there are 210 ways to pick a team of 4 people from a group of 10, which is option B.

53) The average of the numbers is calculated by dividing the sum of the numbers by the count of the numbers. Given that the average of the numbers 7, 12, 25, and y is 15, we can set up the equation:

$$\frac{7+12+25+y}{4} = 15.$$

Simplifying the equation:

$$44 + y = 15 \times 4,$$

$$44 + y = 60,$$

$$y = 60 - 44 = 16.$$

Therefore, the value of y is 16, which is option D.

54) Given that angle D is $45°$ and is represented as $3x - 6°$, we first find the value of x:

$$3x - 6 = 45.$$

Solving for x:

$$3x = 45 + 6 \Rightarrow x = 17.$$

Now, using the value of x to find y, from the angle sum property of a triangle, which states that the sum of angles in a triangle is $180°$:

$$45° + (5 \times 17)° + (y-8)° = 180°.$$

15.3 Answers with Explanation

Simplify and solve for y:

$$45 + 85 + y - 8 = 180 \Rightarrow y = 58.$$

Therefore, the value of y is 58°.

55) Let the angle be x degrees. The complement of an angle is 90 degrees minus the angle. Since the angle is one fifth of its complement, we can set up the equation:

$$x = \frac{1}{5}(90 - x).$$

Solving for x:

$$x = 18 - \frac{1}{5}x \Rightarrow \frac{6}{5}x = 18 \Rightarrow x = \frac{18 \times 5}{6} = 15.$$

Therefore, the measure of the angle is 15 degrees, making option C correct.

56) Let the number be x. According to the problem, when x is subtracted from 35 and the difference is divided by x, the result is 4. This can be written as an equation:

$$\frac{35 - x}{x} = 4.$$

Solving for x:

$$35 - x = 4x \Rightarrow 35 = 5x \Rightarrow x = \frac{35}{5} = 7.$$

Therefore, the value of the number is 7, which is option C.

57) To find the inverse of the function $f(x) = \sqrt{x} + 2$, we first replace $f(x)$ with y:

$$y = \sqrt{x} + 2.$$

To find the inverse, we interchange x and y and solve for y:

$$x = \sqrt{y} + 2.$$

Isolating $\sqrt{y}$:

$$\sqrt{y} = x - 2.$$

Squaring both sides to remove the square root:

$$y = (x-2)^2 = x^2 - 4x + 4.$$

Therefore, the inverse function is $y = x^2 - 4x + 4$, which matches option C.

58) To find the total volume of the solution, we can set up a proportion based on the percentage of alcohol. Since the solution is 8% alcohol, and there are 32 ml of alcohol, we can write the equation:

$$\frac{8}{100} = \frac{32}{\text{Total Volume}}.$$

Solving for the total volume:

$$\text{Total Volume} = \frac{32}{0.08} = 400.$$

Therefore, the total volume of the solution is 400 ml, which is option B.

59) To solve the equation $5\sqrt{2y} + 7 = 27$, first isolate the radical expression:

$$5\sqrt{2y} = 27 - 7 = 20.$$

Divide both sides by 5:

$$\sqrt{2y} = \frac{20}{5} = 4.$$

Square both sides to remove the square root:

$$2y = 4^2 = 16.$$

Finally, solve for y:

$$y = \frac{16}{2} = 8.$$

Therefore, the value of y is 8.

60) To find $k(h(6))$, we first evaluate $h(6)$ and then apply the result to the function k.

First, evaluate $h(6)$:

$$h(6) = 5.$$

15.3 Answers with Explanation

Next, use this result in k:

$$k(h(6)) = k(5) = 7.$$

Therefore, the value of $k(h(6))$ is 7, which is option A.

16. Practice Test 2

16.1 Practices

1) A student has two part-time jobs. Her combined work schedules consist of less than 50 hours per week. Which graph best represents the solution set for all possible combinations of x, the number of hours she worked at her first job, and y, the number of hours she worked at her second job, in one week?

Option A

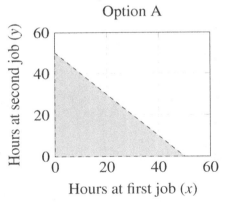

Option B

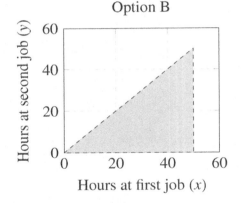

Option C

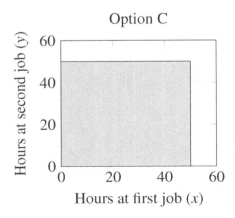

Option D

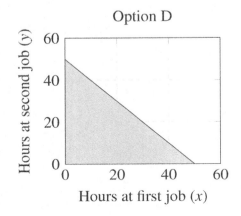

16.1 Practices

2) In the standard (x, y) coordinate plane, which of the following lines contains the points $(4, -10)$ and $(6, 10)$?
   - ☐ A. $y = 5x - 30$
   - ☐ B. $y = -5x + 10$
   - ☐ C. $y = 10x - 50$
   - ☐ D. $y = -10x + 50$

3) Which of the following is equal to the expression below?

$$(3x - 4y)(5x + 2y).$$

   - ☐ A. $15x^2 + 6xy - 8y^2$
   - ☐ B. $15x^2 - 2y^2$
   - ☐ C. $7x^2 + 10xy - 8y^2$
   - ☐ D. $15x^2 - 14xy - 8y^2$

4) Which answer choice best represents the domain and range of the function $y = \sqrt{x+4}$?
   - ☐ A. Domain: $x \geq -4$

     Range: $y \geq 0$
   - ☐ B. Domain: $-4 \leq x \leq 4$

     Range: $y \geq 0$
   - ☐ C. Domain: $y \geq 0$

     Range: $x \geq -4$
   - ☐ D. Domain: $x \leq -4$

     Range: $y \geq 0$

5) What is the slope of a line that is perpendicular to the line $3x + 5y = 7$?
   - ☐ A. $-\frac{3}{5}$
   - ☐ B. $\frac{5}{3}$
   - ☐ C. $-\frac{5}{3}$
   - ☐ D. $\frac{3}{5}$

6) Tickets to a concert cost $10 for adults and $6 for children. A group of 10 people purchased tickets for $80. How many children's tickets did they buy?

- A. 2
- B. 4
- C. 5
- D. 6

7) Which graph best represents part of a quadratic function with a range of all real numbers less than −2?

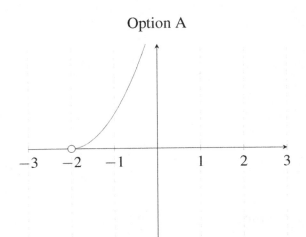
Option A

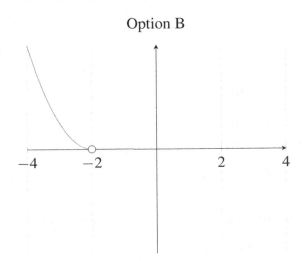
Option B

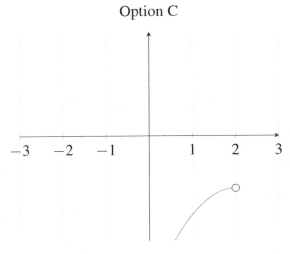
Option C

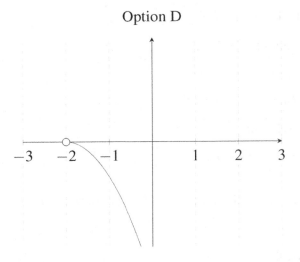
Option D

8) If the function is defined as $f(x) = ax^2 + 10$, where a is a constant, and $f(2) = 30$, what is the value of $f(3)$?
- A. 30
- B. 40
- C. 45

16.1 Practices

☐ D. 55

9) The graph of a line is shown on the grid. The coordinates of both points indicated on the graph of the line are integers.

What is the rate of change of y with respect to x for this line?

☐ A. $\frac{4}{3}$

☐ B. $-\frac{3}{4}$

☐ C. -2

☐ D. -1

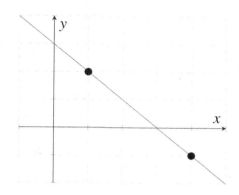

10) What are the equation and slope of the line shown on the grid?

☐ A. $y = 5$; the slope is zero.

☐ B. $x = 5$; the slope is undefined.

☐ C. $y = 5$; the slope is 5.

☐ D. $x = 5$; the slope is 1.

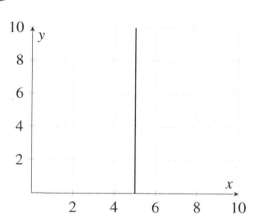

11) In a sequence of numbers, $a_1 = 5$, $a_2 = 9$, $a_3 = 13$, $a_4 = 17$, and $a_5 = 21$. Based on this information, which equation can be used to find the nth term in the sequence, a_n?

☐ A. $a_n = n - 4$

☐ B. $a_n = n + 4$

☐ C. $a_n = 4n - 3$

☐ D. $a_n = 4n + 1$

12) A study tracks the growth of a plant over several weeks. The scatterplot and table show the number of weeks since planting and the height of the plant in centimeters. A linear function can be used to model this relationship.

Which function best models the data?

Weeks Since Planting (x)	Plant Height (cm) (y)
0	15.0
2	20.5
4	24.8
6	30.3
8	35.2

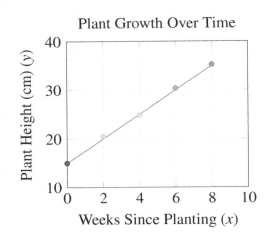

☐ A. $y = 2.5x + 15$

☐ B. $y = -3x + 20$

☐ C. $y = 20x - 3$

☐ D. $y = -3x + 25$

13) Which graph best represents a system of equations that has no solution?

Option A

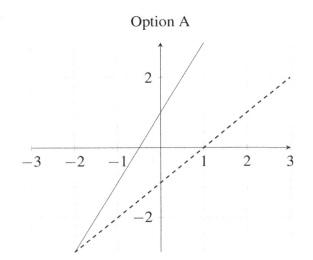

Option B

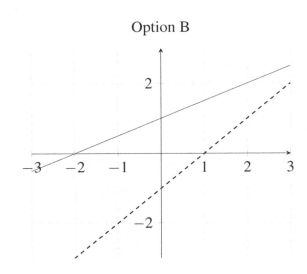

16.1 Practices

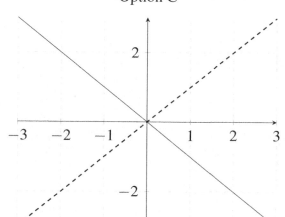
Option C

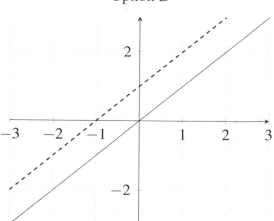
Option D

14) The expression $m^{-2}(m^3)^2$ is equivalent to m^y. What is the value of y? Write your answer in the box.

15) What is the y-intercept of the line with the equation $2x - 5y = 10$?
- ☐ A. −4
- ☐ B. −2
- ☐ C. 2
- ☐ D. 5

16) The tables of ordered pairs represent some points on the graphs of lines L_1 and L_2. Which system of equations is represented by lines L_1 and L_2?

- ☐ A. $\begin{array}{l} 2x - 3y = 9 \\ x - 2y = -4 \end{array}$

- ☐ B. $\begin{array}{l} 3x - 2y = -9 \\ x - 2y = 4 \end{array}$

- ☐ C. $\begin{array}{l} 2x - 3y = 4 \\ x + y = -9 \end{array}$

- ☐ D. $\begin{array}{l} x - 3y = 2 \\ 2x + y = 4 \end{array}$

Line L_1

x	y
1	−1
2	0
5	2

Line L_2

x	y
−8	−1
−9	0
−10	1

17) Simplify $3x^3y^2(2x^4y^3)^2 =$.

- ☐ A. $12x^7y^5$
- ☐ B. $12x^{11}y^8$
- ☐ C. $36x^{11}y^8$
- ☐ D. $36x^7y^5$

18) What are the zeroes of the function $f(x) = x^3 - 6x^2 + 9x$?

- ☐ A. 0
- ☐ B. 3
- ☐ C. 0, 3
- ☐ D. 0, −3

19) Which expression is equivalent to $0.00045 \times (2.5 \times 10^3)$?

- ☐ A. 1.125×10^{-1}
- ☐ B. 1.125
- ☐ C. 1.125×10
- ☐ D. 1.125×10^2

20) What is the positive solution to the equation $2(x+2)^2 = 18 - 11x$. Write your answer in the box.

21) The height of a plant was measured each week for a twelve-week period. The graph shows a linear relationship between the height of the plant in inches and the number of weeks the plant was measured. Which statement best describes the y-intercept of the graph?

16.1 Practices

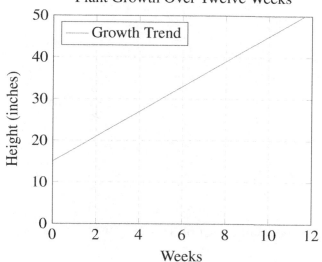

- [ ] A. The height of the plant was measured for 12 weeks.
- [ ] B. The maximum height was 48 inches.
- [ ] C. The height increased by 3 inches per week.
- [ ] D. The height of the plant at the beginning of the measurement period.

22) Find the axis of symmetry of the function $f(x) = -\frac{1}{4}(x+2)^2 + 5$.

- [ ] A. $y = 5$
- [ ] B. There is no axis of symmetry.
- [ ] C. $x = -2$
- [ ] D. $y = -\frac{1}{4}x + 5$

23) The initial value of a computer is $1,200. The value of the computer will decrease at a rate of 15% each year. Which graph best models this situation?

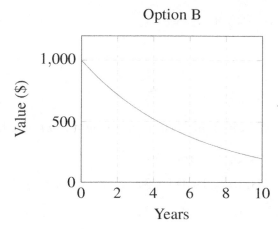

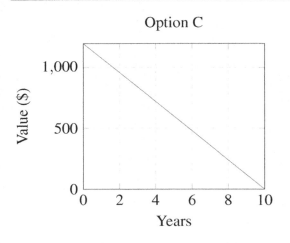
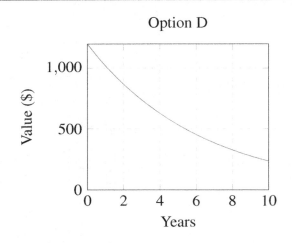

24) Which graph best represents the solution set of $3x + 4y \geq 12$?

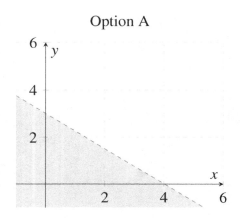
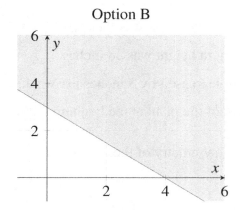
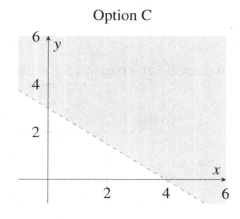
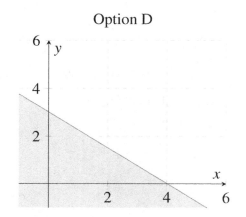

25) Which expression is equivalent to $m^2 - 14m + 45$?

☐ A. $(m+5)(m+9)$

☐ B. $(m-15)(m-3)$

☐ C. $(m+15)(m+3)$

16.1 Practices

☐ D. $(m-5)(m-9)$

26) The table shows the linear relationship between the revenue generated in thousand dollars by a bookshop and the number of books sold. What is the rate of change in revenue in thousand dollars with respect to the number of books sold in the shop?

☐ A. 0.04

☐ B. 0.06

☐ C. 0.02

☐ D. 0.08

Books Sold	Revenue (in thousand dollars)
50	2
100	4
150	6
200	8

27) Quadratic function f models the height in feet of a ball thrown into the air t seconds after it is thrown. The graph of the function is shown. What is the maximum value of the graph of the function? Write your answer in the box: ☐

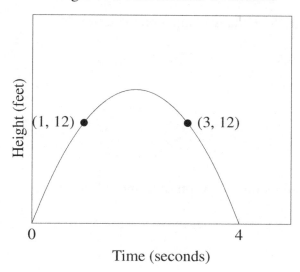

Height of a Thrown Ball Over Time

28) What is the value of z in the following system of equations?

$$\begin{cases} 4z + 3y = 6 \\ y = z \end{cases}$$

- A. $z = \frac{1}{2}$
- B. $z = \frac{3}{4}$
- C. $z = \frac{6}{7}$
- D. $z = \frac{5}{4}$

29) The graphs of linear functions h and k are displayed on the grid. Which function is best represented by the graph of k?

- A. $k(x) = 2h(x) + 5$
- B. $k(x) = \frac{1}{2}h(x) - 5$
- C. $k(x) = 2h(x) - 5$
- D. $k(x) = \frac{1}{2}h(x) + 5$

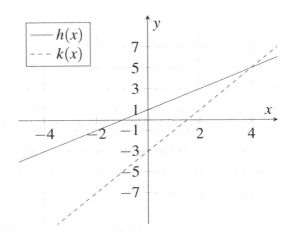

30) The equation $y^2 = 5y - 4$ has how many distinct real solutions?

- A. 0
- B. 1
- C. 2
- D. 3

31) A table represents some points on the graph of a linear function. Which equation represents the same relationship?

- A. $y - 5 = 3(x + 20)$
- B. $y - 20 = \frac{1}{3}(x + 5)$
- C. $y + 5 = \frac{1}{3}(x - 20)$
- D. $y + 20 = 3(x - 5)$

x	y
−5	20
0	$\frac{65}{3}$
15	$\frac{80}{3}$

32) Examine the graph of a linear equation provided below.

What are the coordinates of the x-intercept in this graph?

16.1 Practices

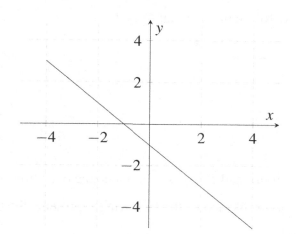

- [ ] A. $(0,-1)$
- [ ] B. $(0,-4)$
- [ ] C. $(-1,0)$
- [ ] D. $(-3,0)$

33) An exponential function is depicted on the provided grid. Which function most accurately represents the graph?

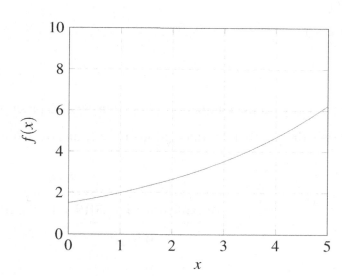

- [ ] A. $f(x) = 3(0.33)^x$
- [ ] B. $f(x) = 3(0.67)^x$
- [ ] C. $f(x) = 1.5(1.33)^x$
- [ ] D. $f(x) = 1.5(33)^x$

34) Given $x = 7$, calculate the value of y in the equation:

$$2y = \frac{3x^2}{5} - 5.$$

Write your answer in the box: ☐.

35) Determine the domain of the function $g(x) = -5x^2 + 36$.

- [ ] A. $(-\infty, 36]$
- [ ] B. $(-6, 6)$
- [ ] C. $[-3, 3]$
- [ ] D. $\mathbb{R}$

36) Which of the following is equal to $c^{\frac{4}{7}}$?

- [ ] A. $c^{\frac{7}{4}}$
- [ ] B. $\sqrt{c^{\frac{7}{4}}}$
- [ ] C. $\sqrt[7]{c^4}$
- [ ] D. $\sqrt[4]{c^7}$

37) On Sunday, John read P pages of a magazine each hour for 5 hours, and Lisa read Q pages of a magazine each hour for 2 hours. Which of the following represents the total number of magazine pages read by John and Lisa on Sunday?

- [ ] A. $7PQ$
- [ ] B. $10PQ$
- [ ] C. $5P + 2Q$
- [ ] D. $2P + 5Q$

38) A business analyzed the number of online transactions processed through their website since January 2019. The table shows the number of transactions in millions over time. This data can be modeled by a quadratic function.

Months Since Jan 2019	Transactions (millions)
0	5.0
6	5.0
12	41.0
18	113.0
24	221.0

Which function best models this data?

- [ ] A. $f(x) = 0.5x^2 - 3x + 5$
- [ ] B. $f(x) = -0.5x^2 + 3x - 5$
- [ ] C. $f(x) = 0.5x^2 + 3x + 5$
- [ ] D. $f(x) = x^2 - 6x + 10$

39) Which of the following represents the graph of the line with the equation $3x + 4y = 12$?

Option A
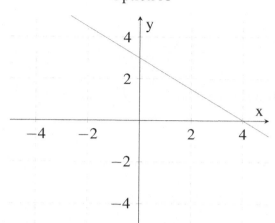

Option B
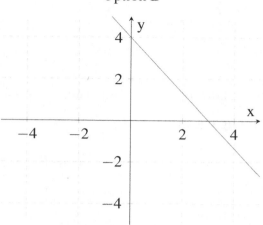

Option C
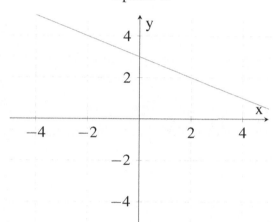

Option D
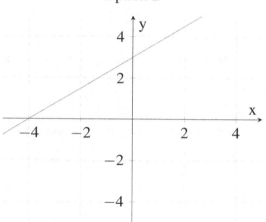

40) Determine the solution (x, y) for the following system of equations:

$$3x + 4y = 10$$
$$5x - 8y = -15$$

- [ ] A. $\left(\frac{5}{8}, \frac{9}{4}\right)$
- [ ] B. $\left(\frac{3}{8}, \frac{5}{8}\right)$
- [ ] C. $\left(\frac{5}{11}, \frac{95}{44}\right)$
- [ ] D. $\left(\frac{3}{11}, \frac{5}{44}\right)$

41) Find the solutions for the equation $4x^2 - 5x = 4 - x$.

- [ ] A. $x = 4$ and $x = 1$

B. $x = 4$ and $x = -\frac{4}{5}$

C. $x = \frac{2+\sqrt{5}}{4}$ and $x = \frac{2-\sqrt{5}}{4}$

D. $x = \frac{1+\sqrt{5}}{2}$ and $x = \frac{1-\sqrt{5}}{2}$

42) The table details the linear relationship between the amount of water in a tank (in gallons) and the time in hours since filling began.

Time (hours)	Water (gallons)
0	0
2	50
4	100
6	150
8	200

Based on the table, what is the rate of change of the water level in the tank in gallons per hour? Write your answer in the box. [ ]

43) The function $h(x)$ is described by a polynomial. Some values of x and $h(x)$ are listed in the following table. Which of the following must be a factor of $h(x)$?

A. $x+2$

B. $x-3$

C. $x+3$

D. $x-4$

x	$h(x)$
0	3
1	0
2	−1
3	0
4	3

44) Determine the value of $\frac{5d}{e}$ when $\frac{e}{d} = 3$.

A. 0

B. $\frac{3}{5}$

C. $\frac{5}{3}$

D. 3

45) Which of the following equations represents a graph that is a straight line?

16.1 Practices

- [ ] A. $y = 2x^2 + 8$
- [ ] B. $x^2 + 4y^2 = 4$
- [ ] C. $5x - 3y = 5x$
- [ ] D. $8x + 3xy = 9$

46) If $\frac{c-d}{d} = \frac{12}{15}$, then which of the following must be true?

- [ ] A. $\frac{c}{d} = \frac{12}{15}$
- [ ] B. $\frac{c}{d} = \frac{12}{27}$
- [ ] C. $\frac{c}{d} = \frac{27}{15}$
- [ ] D. $\frac{c}{d} = \frac{24}{12}$

47) Simplify the expression $3\sqrt{18} + 3\sqrt{3}$.

- [ ] A. $12\sqrt{5}$
- [ ] B. $6\sqrt{21}$
- [ ] C. $9\sqrt{2} + 3\sqrt{3}$
- [ ] D. $6\sqrt{3}$

48) Identify which table shows y as a function of x.

Table A:

x	y
1	2
2	4
2	4
4	8

Table B:

x	y
1	2
2	4
2	5
4	8

Table C:

x	y
1	2
2	4
4	6
4	8

Table D:

x	y
1	2
3	6
3	8
3	8

49) The graph of a quadratic function is displayed on a grid.

Which equation best represents the axis of symmetry for this graph?

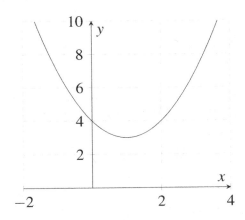

- A. $y = 8$
- B. $x = 1$
- C. $y = 4$
- D. $x = -2$

50) A tree increases in height at a constant rate. After six weeks, the tree is 36 *cm* tall. Which of the following functions represents the relationship between the height *h* of the tree and the number of weeks *w* since it was measured?

- A. $h(w) = 36w + 6$
- B. $h(w) = 6w + 36$
- C. $h(w) = 36w$
- D. $h(w) = 6w$

51) If $a \boxtimes b = \sqrt{a^2 - b}$, what is the value of $7 \boxtimes 21$?

- A. $2\sqrt{7}$
- B. $\sqrt{23}$
- C. $3\sqrt{14}$
- D. $\sqrt{17}$

52) Calculate the product of all possible values of *y* in the equation $|y - 8| = 5$.

- A. 2
- B. 8
- C. 11
- D. 39

53) The average of seven numbers is 32. If an eighth number, 50, is added, what is the new average? (Round your answer to the nearest hundredth.)

- A. 31

16.1 Practices

☐ B. 33.38

☐ C. 34.25

☐ D. 35.75

54) A credit union offers 3.75% simple interest on a savings account. If you deposit $15,000, how much interest will you earn in three years?

☐ A. $562.50

☐ B. $1,687.50

☐ C. $3,375

☐ D. $4,500

55) If $h(x) = 3x^3 + 4x^2 - x$ and $k(x) = -3$, what is the value of $h(k(x))$?

☐ A. -3

☐ B. -39

☐ C. 26

☐ D. -42

56) If 200% of a number is 100, then what is 60% of that number?

☐ A. 30

☐ B. 40

☐ C. 50

☐ D. 60

57) 10 students had an average score of 80. The remaining 8 students of the class had an average score of 90. What is approximately the mean (average) score of the entire class?

☐ A. 83.3

☐ B. 84.4

☐ C. 84.5

☐ D. 86.7

58) John orders a pack of notebooks for $4 per pack. A tax of 7.5% is added to the cost of the notebooks before a flat shipping fee of $5 rounds out the transaction. Which of the following represents the total cost of n packs of notebooks in dollars?

☐ A. $4.75n+5$

☐ B. $4n+4$

☐ C. $11.5n+4$

☐ D. $4.3n+5$

59) If $z-4=5$ and $3w+2=8$, what is the value of $zw+10$?

☐ A. 28

☐ B. 22

☐ C. 19

☐ D. 15

60) The following graph represents a polynomial function. What is the number of roots of this function?

☐ A. 0

☐ B. 1

☐ C. 2

☐ D. 3

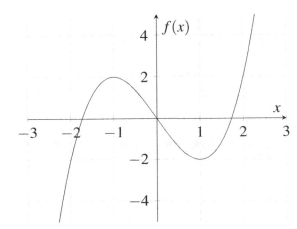

16.2 Answer Keys

1) A. Option A
2) C. $y = 10x - 50$
3) D. $15x^2 - 14xy - 8y^2$
4) A. Domain: $x \geq -4$
   Range: $y \geq 0$
5) B. $\frac{5}{3}$
6) C. 5
7) C. Option C
8) D. 55
9) D. -1
10) B. $x = 5$; the slope is undefined.
11) D. $a_n = 4n + 1$
12) A. $y = 2.5x + 15$
13) Option D.
14) 4
15) B. -2
16) C. $\begin{array}{l} 2x - 3y = 4 \\ x + y = -9 \end{array}$
17) B. $12x^{11}y^8$
18) C. 0, 3
19) B. 1.125
20) $\frac{1}{2}$
21) D.
22) C. $x = -2$
23) Option D.
24) Option B.
25) D. $(m-5)(m-9)$
26) A. 0.04
27) 16 feet
28) C. $z = \frac{6}{7}$
29) C. $k(x) = 2h(x) - 5$
30) C. 2
31) B. $y - 20 = \frac{1}{3}(x + 5)$
32) C. $(-1, 0)$
33) C. $f(x) = 1.5(1.33)^x$
34) 12.2
35) D. $\mathbb{R}$
36) C. $\sqrt[7]{c^4}$
37) C. $5P + 2Q$
38) A. $f(x) = 0.5x^2 - 3x + 5$
39) A. Option A
40) C. $\left(\frac{5}{11}, \frac{95}{44}\right)$
41) D. $x = \frac{1+\sqrt{5}}{2}$ and $x = \frac{1-\sqrt{5}}{2}$
42) 25 gallons per hour
43) B. $x - 3$
44) C. $\frac{5}{3}$
45) C. $5x - 3y = 5x$
46) C. $\frac{c}{d} = \frac{27}{15}$
47) C. $9\sqrt{2} + 3\sqrt{3}$
48) Table A
49) B. $x = 1$
50) D. $h(w) = 6w$
51) A. $2\sqrt{7}$
52) D. 39
53) C. 34.25
54) B. $1,687.50

55) D. −42
56) A. 30
57) B. 84.4
58) D. 4.3n + 5
59) A. 28
60) D. 3

16.3 Answers with Explanation

1) The correct graph must represent all combinations of x and y where the total hours worked $(x+y)$ are less than 50. In Option A, the shaded area below the dashed line $y = 50 - x$ shows all points where the sum of x and y is less than 50, correctly representing the constraints of the problem. The other options either represent incorrect relationships or do not correctly depict the total hours constraint. In option D, the line is solid which is incorrect.

2) To find the equation of the line that passes through $(4, -10)$ and $(6, 10)$, we first calculate the slope:

$$\text{Slope} = \frac{10 - (-10)}{6 - 4} = \frac{20}{2} = 10.$$

Using the point-slope form of a line, $y - y_1 = m(x - x_1)$, and the point $(4, -10)$:

$$y - (-10) = 10(x - 4) \Rightarrow y + 10 = 10x - 40 \Rightarrow y = 10x - 40 - 10 \Rightarrow y = 10x - 50.$$

Therefore, the equation of the line is $y = 10x - 50$, which corresponds to option C.

3) Multiplying the polynomials $(3x - 4y)(5x + 2y)$ using the distributive property (FOIL method):

$$3x \times 5x = 15x^2, \quad 3x \times 2y = 6xy, \quad -4y \times 5x = -20xy, \quad -4y \times 2y = -8y^2.$$

Adding these together:

$$15x^2 + 6xy - 20xy - 8y^2 = 15x^2 - 14xy - 8y^2.$$

So, the correct answer is option D.

4) For the function $y = \sqrt{x+4}$, the domain and range can be determined as follows:

- Domain: Since the square root function is defined for non-negative numbers, $x + 4 \geq 0$. Therefore, $x \geq -4$.

- Range: The square root function always yields non-negative values. Thus, $y \geq 0$.

Therefore, the correct answer is option A, with the domain $x \geq -4$ and the range $y \geq 0$.

5) The slope of the line $3x+5y=7$ can be found by rearranging it into slope-intercept form $y = mx+b$:

$$5y = -3x+7,$$

$$y = -\frac{3}{5}x + \frac{7}{5}.$$

The slope of this line is $-\frac{3}{5}$. The slope of a line perpendicular to this line is the negative reciprocal of $-\frac{3}{5}$, which is $\frac{5}{3}$.

Therefore, the correct answer is option B, $\frac{5}{3}$.

6) Let the number of children's tickets be c and the number of adult tickets be $10-c$. The total cost is $80. Setting up the equation:

$$10(10-c) + 6c = 80.$$

Solving for c:

$$100 - 10c + 6c = 80 \Rightarrow -4c = -20 \Rightarrow c = 5.$$

Therefore, the number of children's tickets bought is 5, which is option C.

7) The correct graph must represent a quadratic function with a range of all real numbers less than -2.

In Options A and B, the range is $y > 0$.

In Option D, the range is $y < 0$.

In Option C, the graph correctly models the condition that the function's range is all real numbers less than -2.

8) First, find the value of a using $f(2) = 30$:

$$30 = a(2)^2 + 10 \Rightarrow 30 = 4a + 10 \Rightarrow 4a = 20 \Rightarrow a = 5.$$

Now, find $f(3)$ using $a = 5$:

$$f(3) = 5(3)^2 + 10,$$

$$f(3) = 5 \times 9 + 10 = 45 + 10 = 55.$$

Therefore, the value of $f(3)$ is 55, which is option D.

16.3 Answers with Explanation

9) The rate of change (slope) of a line is calculated using two points on the line. For the points $(1,2)$ and $(4,-1)$, the slope is calculated as the difference in y-coordinates divided by the difference in x-coordinates:

$$\text{Slope} = \frac{-1-2}{4-1} = -1.$$

This slope indicates that for each unit increase in x, y decreases by 1 unit.

Therefore, the correct answer is option D.

10) The graph shows a vertical line crossing the x-axis at $x = 5$. For a vertical line, the x-coordinate is constant while the y-coordinate changes. This line is described by the equation $x = 5$.

The slope of a vertical line is undefined because the change in y is not zero while the change in x is zero, leading to a division by zero when calculating the slope. Thus, the correct answer is option B.

11) To find the nth term of the sequence, observe the pattern: each term increases by 4. The first term is 5, the second is $9 = (5+4)$, the third is $13 = (9+4)$, and so on. The formula for the nth term can be found by noticing that each term is four times the term number plus 1:

$$a_n = 4n + 1.$$

Therefore, the correct equation for the nth term in this sequence is option D, $a_n = 4n + 1$.

12) The scatter plot and table show a linear relationship between the number of weeks since planting and the height of the plant. The function $y = 2.5x + 15$ approximately models this relationship. In this function, the slope of 2.5 indicates that the plant grows by 2.5 cm each week, and the y-intercept of 15 represents the initial height of the plant at the time of planting (0 weeks). The other options either have incorrect slopes or y-intercepts that do not align with the data presented.

13) A system of linear equations has no solution when the lines represented by the equations are parallel and never intersect. Among the options, Option D is the only one that shows two parallel lines. Since these lines never meet, there is no point that satisfies both equations simultaneously, indicating that the system has no solution. The other options depict intersecting lines, suggesting systems with one solution.

14) To simplify the expression $m^{-2}(m^3)^2$ and find y:

$$m^{-2}(m^3)^2 = m^{-2} \cdot m^6 = m^{-2+6} = m^4.$$

Therefore, the expression is equivalent to m^4, and the value of y is 4.

15) The y-intercept of a line occurs where $x = 0$. Substituting $x = 0$ into the equation $2x - 5y = 10$:

$$2(0) - 5y = 10 \Rightarrow -5y = 10 \Rightarrow y = -2.$$

Therefore, the y-intercept is -2, which is option B.

16) The tables for lines L_1 and L_2 show points that satisfy the equations $2x - 3y = 4$ and $x + y = -9$, respectively. These equations correspond to Option C. The points in the table for L_1 fit the equation $2x - 3y = 4$, and the points in the table for L_2 fit the equation $x + y = -9$. This makes Option C the correct choice, as it represents the system of equations depicted by the tables.

17) To simplify the expression $3x^3y^2\left(2x^4y^3\right)^2$:

$$3x^3y^2 \cdot (2x^4y^3)^2 = 3x^3y^2 \cdot 4x^8y^6 = 12x^{11}y^8.$$

Therefore, the simplified expression is $12x^{11}y^8$, which is option B.

18) To find the zeroes of $f(x) = x^3 - 6x^2 + 9x$, factor out the common term:

$$f(x) = x(x^2 - 6x + 9).$$

The quadratic can be factored as:

$$f(x) = x(x-3)^2.$$

Setting each factor to zero gives the zeroes:

$$x = 0, \; x - 3 = 0 \Rightarrow x = 3.$$

Therefore, the zeroes are $0, 3$, which is option C.

19) To simplify $0.00045 \times (2.5 \times 10^3)$:

$$0.00045 \times 2.5 \times 10^3 = 4.5 \times 10^{-4} \times 2.5 \times 10^3 = 11.25 \times 10^{-4+3} = 11.25 \times 10^{-1}.$$

Converting 11.25×10^{-1} back to decimal form:

$$11.25 \times \frac{1}{10} = 1.125.$$

Therefore, the equivalent expression is 1.125, which is option B.

20) To solve the equation $2(x+2)^2 = 18 - 11x$:

$$2(x+2)^2 + 11x = 18 \Rightarrow 2(x^2 + 4x + 4) + 11x = 18$$
$$\Rightarrow 2x^2 + 19x + 8 = 18$$
$$\Rightarrow 2x^2 + 19x - 10 = 0.$$

By factoring we have:

$$(2x - 1)(x + 10) = 0.$$

The positive solution:

$$2x - 1 = 0 \Rightarrow x = \frac{1}{2}.$$

Therefore, the positive solution of the equation is $x = \frac{1}{2}$.

21) The y-intercept of the graph represents the height of the plant at the beginning of the measurement period (week 0), which is 15 inches in this case. It does not reflect the duration of the measurement, the maximum height, or the final height. The slope of the line, which is 3, indicates that the height of the plant increased by 3 inches per week. Therefore, the best description of the y-intercept is provided by Option D.

22) The axis of symmetry of a quadratic function $f(x) = a(x - h)^2 + k$ is given by the line $x = h$. For the function $f(x) = -\frac{1}{4}(x+2)^2 + 5$, the form matches $a(x+2)^2 + k$, where $h = -2$.

Therefore, the axis of symmetry is $x = -2$, which is option C.

23) The correct graph for this scenario is Option D, which shows an exponential decay starting at $1,200 and decreasing by 15% each year. The graph depicts the value as continuously decreasing, but the rate of decrease slows down over time, which is characteristic of exponential decay. The value approaches but never reaches zero, accurately representing the depreciation of the computer over time.

24) The inequality $3x + 4y \geq 12$ is represented by a line and the region above it. Option B correctly uses a solid line to indicate that points on the line are included in the solution set (the equality part of the inequality). The shaded area above the line in Option B shows the region where $3x + 4y$ is greater than 12. This graph accurately represents the solution set of the given inequality.

25) To factorize $m^2 - 14m + 45$, look for two numbers that multiply to 45 and add up to -14. The numbers -5 and -9 fit this criterion:
$$m^2 - 14m + 45 = (m-5)(m-9).$$

Therefore, the equivalent expression is $(m-5)(m-9)$, which is option D.

26) The rate of change in revenue can be determined by calculating the slope of the line representing the relationship between the number of books sold and the revenue generated. From the table, the slope is calculated as the change in revenue divided by the change in the number of books sold. The change in revenue is $6,000 (from $2,000 to $8,000), and the change in books sold is 150 (from 50 to 200 books). The slope is therefore:

$$\frac{\$6 \text{ thousand dollars}}{150 \text{ books}} = \$0.04 \text{ thousand dollars per book}.$$

This represents an increase in revenue of $0.04 thousand dollars for each additional book sold, making Option A the correct answer.

27) By solving the system of equations derived from the given points (1, 12), (3, 12), and (4, 0) for the quadratic function $f(x) = ax^2 + bx + c$ (with $c = 0$ as the graph passes through (0, 0)), we find that the maximum value of the function f at $x = 2$ is 16. The graph of quadratic function $f(x) = ax^2 + bx + c$ passes through $(0, 0)$. Thus, $f(0) = 0$ and hence $c = 0$.

16.3 Answers with Explanation

We see that the graph also passes through points $(1, 12)$, $(3, 12)$, and $(4, 0)$. Therefore, we have:

$$f(1) = 12 \Rightarrow a(1)^2 + b(1) = 12 \Rightarrow a + b = 12$$
$$f(3) = 12 \Rightarrow a(3)^2 + b(3) = 12 \Rightarrow 9a + 3b = 12.$$

By solving the above system of equations, we find $a = -4$ and $b = 16$. Hence:

$$f(x) = -4x^2 + 16x.$$

To find the maximum value of f, which occurs at the vertex of the parabola, we evaluate $f(2)$ as the vertex lies midway between $x = 1$ and $x = 3$ (the symmetry of the parabola):

$$\text{Maximum value} = f(2) = -4(2)^2 + 16(2) = -16 + 32 = 16.$$

28) Substituting $y = z$ into the first equation:

$$4z + 3z = 6 \Rightarrow 7z = 6 \Rightarrow z = \frac{6}{7}.$$

Therefore, the value of z is $\frac{6}{7}$, which is option C.

29) We can consider two points of each graph and calculate the slpoe of each line. In the graph, h is a linear function with a slope of 1. The slpoe of function k is 2. Thus,

$$k(x) = 2h(x) + a,$$

where a is a constant. Since $h(4) = k(4) = 5$, then we get:

$$5 = 2(5) + a \Rightarrow a = -5.$$

Therefore, $k(x) = 2h(x) - 5$, making option C correct.

30) Rewrite the equation in standard quadratic form:

$$y^2 - 5y + 4 = 0.$$

Factor this equation, we obtain:

$$(y-4)(y-1) = 0.$$

The solutions are $y = 4$ and $y = 1$, which are distinct real solutions.

Therefore, the equation has 2 distinct real solutions, which is option C.

31) The table provides the points $(-5, 20)$, $(0, \frac{65}{3})$, and $(15, \frac{80}{3})$ that lie on the graph of a linear function. These points align with the equation $y = \frac{1}{3}x + \frac{65}{3}$, which is the simplified form of Option B. The relationship between x and y in these points indicates a slope of $\frac{1}{3}$ and a y-intercept of $\frac{65}{3}$. This matches the equation in Option B, making it the correct choice for representing the same relationship.

32) The x-intercept of a graph is the point where the line crosses the x-axis, which occurs where the y-coordinate is zero. In the provided graph, the line intersects the x-axis at the point $(-1, 0)$. Therefore, the coordinates of the x-intercept are $(-1, 0)$, which corresponds to Option C.

33) The provided graph shows an exponential increase, which is consistent with the function in Option C, $f(x) = 1.5(1.33)^x$. This function starts with a base value of 1.5 and grows at a rate of 1.33 per x increment, indicating an exponential growth pattern. The other options either decrease exponentially (Options A and B) or increase at a significantly different rate (Option D), which does not match the graph's pattern. Therefore, Option C is the most accurate representation of the graph.

34) Substituting $x = 7$ into the equation:

$$2y = \frac{3 \times 7^2}{5} - 5 = \frac{3 \times 49}{5} - 5 = \frac{147}{5} - 5 = 29.4 - 5 = 24.4.$$

Solving for y:

$$y = \frac{24.4}{2} = 12.2.$$

35) Since $g(x) = -5x^2 + 36$ is a quadratic function, it is defined for all real numbers. Therefore, the domain

16.3 Answers with Explanation

is all real numbers, $\mathbb{R}$.

36) $c^{\frac{4}{7}}$ is equivalent to the 7-th root of c^4, which is $\sqrt[7]{c^4}$.

37) John reads $5P$ pages in total (5 hours times P pages per hour), and Lisa reads $2Q$ pages in total (2 hours times Q pages per hour). The total number of pages read is $5P + 2Q$.

38) The correct equation is identified by examining the trend of the data in the table and comparing it to the equations of the given options. The table shows a quadratic trend with an increasing rate of transactions over time. Option A, $f(x) = 0.5x^2 - 3x + 5$, closely aligns with this trend, as it correctly models the increase in the number of transactions over the months. The coefficients and constants in this equation match the pattern observed in the data, making it the best choice for representing this relationship.

39) To find the correct graph for the line $3x + 4y = 12$, the equation is first rearranged into slope-intercept form, which is $y = mx + b$. After rearrangement, the equation becomes $y = -\frac{3}{4}x + 3$. Option A correctly shows this line, with a slope of $-\frac{3}{4}$ and a y-intercept of 3. The other options either have incorrect slopes or incorrect y-intercepts, making them incorrect representations of the line.

40) Solving the system of equations:

From $3x + 4y = 10$:

$$4y = 10 - 3x \Rightarrow y = \frac{10 - 3x}{4}.$$

Substitute y in $5x - 8y = -15$:

$$5x - 8\left(\frac{10-3x}{4}\right) = -15 \Rightarrow 5x - 2(10-3x) = -15 \Rightarrow 5x + 6x - 20 = -15 \Rightarrow 11x = 5.$$

Thus, $x = \frac{5}{11}$. Now, we can find y:

$$y = \frac{10 - 3\left(\frac{5}{11}\right)}{4} = \frac{1}{4}\left(\frac{95}{11}\right) = \frac{95}{44}.$$

Therefore, the point $\left(\frac{5}{11}, \frac{95}{44}\right)$ is the solution, option C.

41) Rearranging and simplifying the equation:

$$4x^2 - 5x - 4 + x = 0 \Rightarrow 4x^2 - 4x - 4 = 0 \Rightarrow x^2 - x - 1 = 0.$$

Using the quadratic formula:

$$x = \frac{-b \pm \sqrt{b^2 - 4ac}}{2a}$$

$$x = \frac{1 \pm \sqrt{(-1)^2 - 4 \cdot 1 \cdot (-1)}}{2 \cdot 1}$$

$$x = \frac{1 \pm \sqrt{1 + 4}}{2}$$

$$x = \frac{1 \pm \sqrt{5}}{2}.$$

Thus, the solutions are $x = \frac{1+\sqrt{5}}{2}$ and $x = \frac{1-\sqrt{5}}{2}$.

42) The rate of change of the water level can be determined by calculating the slope of the line representing the relationship between the water level and time. The slope is the change in the water level divided by the change in time. From the table, the water level increases by 200 gallons over 8 hours, resulting in a rate of change of $\frac{200 \text{ gallons}}{8 \text{ hours}} = 25$ gallons per hour. This indicates that the water level in the tank increases by 25 gallons for each hour.

43) To determine the correct factors of the polynomial $h(x)$, we look for values of x such that $h(x) = 0$. In the given table, $h(x)$ equals 0 at $x = 1$ and $x = 3$. This indicates that $x - 1$ and $x - 3$ must be factors of $h(x)$, as the polynomial function equals zero when x is 1 or 3. Therefore, Option B, $x - 3$, is the correct answer.

44) Given $\frac{e}{d} = 3$, we can rearrange it to $e = 3d$. Substituting this into $\frac{5d}{e}$:

$$\frac{5d}{3d} = \frac{5}{3}.$$

Therefore, $\frac{5d}{e} = \frac{5}{3}$.

45) The equation $5x - 3y = 5x$ simplifies to $-3y = 0$ or $y = 0$, which is a horizontal line. Thus, it represents a graph that is a straight line.

16.3 Answers with Explanation

46) Given $\frac{c-d}{d} = \frac{12}{15}$, we can rewrite it as:

$$\frac{c}{d} - 1 = \frac{12}{15}.$$

Adding 1 to both sides:

$$\frac{c}{d} = \frac{12}{15} + 1 = \frac{12}{15} + \frac{15}{15} = \frac{27}{15}.$$

Therefore, $\frac{c}{d} = \frac{27}{15}$.

47) Simplify the expression:

$$3\sqrt{18} + 3\sqrt{3} = 3\sqrt{9 \times 2} + 3\sqrt{3} = 3 \times 3\sqrt{2} + 3\sqrt{3} = 9\sqrt{2} + 3\sqrt{3}.$$

Since $\sqrt{2}$ and $\sqrt{3}$ cannot be simplified further and are not like terms, the expression is already in its simplest form.

48) A function is defined as a relationship where each input (x) has exactly one output (y). In Option A, every x-value corresponds to one unique y-value, which meets the definition of a function. The other options have duplicate x-values with different y-values, which is not characteristic of a function. Therefore, Option A correctly shows y as a function of x.

49) The axis of symmetry of a quadratic function is a vertical line that passes through the vertex of the parabola. In this graph, the vertex appears to be near $x = 1$. Therefore, the axis of symmetry would be the line $x = 1$. Option B, $x = 1$, is the correct answer as it best represents the axis of symmetry of the graph.

50) The tree grows 36 *cm* over 6 weeks, so the rate of growth per week is $\frac{36\ cm}{6\ weeks} = 6\ cm/week$. Therefore, the height after w weeks can be represented by the function $h(w) = 6w$.

51) Using the custom operation ⊠:

$$7 \boxtimes 21 = \sqrt{7^2 - 21} = \sqrt{49 - 21} = \sqrt{28} = \sqrt{4 \times 7} = 2\sqrt{7}.$$

Therefore, the correct answer is option A.

52) Solve the equation $|y-8| = 5$ for y:

Case 1: $y - 8 = 5$ gives $y = 13$.

Case 2: $y - 8 = -5$ gives $y = 3$.

The product of the two solutions is $13 \times 3 = 39$.

53) The total sum of the first seven numbers is $7 \times 32 = 224$. Adding the eighth number, 50, gives a new total of $224 + 50 = 274$. The new average for eight numbers is $\frac{274}{8} = 34.25$.

54) Simple interest can be calculated using the formula $I = P \times r \times t$, where I is the interest, P is the principal amount, r is the rate of interest, and t is the time in years.

For a deposit of $\$15,000$ at a rate of 3.75% for 3 years:

$$I = \$15,000 \times 3.75\% \times 3 = \$15,000 \times 0.0375 \times 3 = \$1,687.50.$$

55) First, substitute $k(x) = -3$ into $h(x)$:

$$h(k(x)) = h(-3) = 3(-3)^3 + 4(-3)^2 - (-3).$$

Calculate the value:

$$= 3(-27) + 4(9) + 3 = -81 + 36 + 3 = -42.$$

Therefore, the correct answer is option D.

56) First, find the number. Since 200% of the number is 100:

$$\frac{200}{100} \times \text{number} = 100 \Rightarrow \text{number} = \frac{100}{2} = 50.$$

Next, calculate 60% of 50:

$$\frac{60}{100} \times 50 = 30.$$

Therefore, the number is 30, making option A correct.

16.3 Answers with Explanation

57) Calculate the total score for both groups:

10 students × 80 score/student = 800 total score.

8 students × 90 score/student = 720 total score.

The total number of students is $10 + 8 = 18$, and the total score is $800 + 720 = 1520$. The average score is:

$$\frac{1520}{18} \approx 84.44.$$

Rounded to the nearest tenth, this is approximately 84.4.

58) The cost of n packs is $4n$. With 7.5% tax, the total becomes $4n \times 1.075 = 4.3n$. Adding the flat shipping fee of \$5 gives the total cost as $4.3n + 5$.

59) First, solve for z and w: $z - 4 = 5$ gives $z = 9$. $3w + 2 = 8$ gives $3w = 6$ and $w = 2$.

Now, calculate $zw + 10$: $9 \times 2 + 10 = 18 + 10 = 28$, which is option A.

60) The roots of a polynomial function are the x-values where the graph intersects the x-axis. In the provided graph, the polynomial function intersects the x-axis at three distinct points. Therefore, it has three roots, making Option D, 3 roots, the correct answer.

Author's Final Note

I hope you enjoyed this book as much as I enjoyed writing it. I have tried to make it as easy to understand as possible. I have also tried to make it fun. I hope I have succeeded. If you have any suggestions for improvement, please let me know. I would love to hear from you.

The accuracy of examples and practice is very important to me. We have done our best. But I also expect that I have made some minor errors. Constant improvement is the name of the game. If you find any errors, please let me know. I will fix them in the next edition.

Your learning journey does not end here. I have written a series of books to help you learn math. Make sure you browse through them. I especially recommend workbooks and practice tests to help you prepare for your exams.

I also enjoy reading your reviews. If you have a moment, please leave a review on Amazon. It will help other students find this book.

If you have any questions or comments, please feel free to contact me at drNazari@effortlessmath.com.

And one last thing: Remember to use online resources for additional help. I recommend using the resources on `https://effortlessmath.com`. There are many great videos on YouTube.

Good luck with your studies!

Dr. Abolfazl Nazari

Made in United States
North Haven, CT
14 June 2024